~~John Read Middle School~~
~~486 Redding Road~~
~~Redding, Connecticut 06896~~

DOUBLE STARS

the story of

CAROLINE HERSCHEL

DOUBLE STARS

the story of

CAROLINE HERSCHEL

Padma Venkatraman

MORGAN REYNOLDS

PUBLISHING

Greensboro, North Carolina

Profiles

IN SCIENCE

Robert Boyle

Rosalind Franklin

Ibn al-Haytham

Edmond Halley

Marie Curie

Caroline Herschel

DOUBLE STARS: THE STORY OF CAROLINE HERSCHEL

Copyright © 2007 by Padma Venkatraman

Library of Congress Cataloging-in-Publication Data

Venkatraman, Padma.
  Double stars : the story of Caroline Herschel / by Padma Venkatraman.
    p. cm. -- (Profiles in science)
  Includes bibliographical references and index.
  ISBN-13: 978-1-59935-042-4 (alk. paper)
  ISBN-10: 1-59935-042-4 (alk. paper)
  1. Herschel, Caroline Lucretia, 1750-1848. 2. Herschel, William, Sir,
1738-1822. 3. Astronomers--Great Britain--Biography. I. Title.
  QB36.H59V46 2007
  520.92--dc22
  [B]
                                    2006035171

Printed in the United States of America
First Edition

This book is dedicated to R. Lohmann, for his support and encouragement.

Thanks also to Dr. David Evans, Dr. Elizabeth Bozyan, and A. Venkatraman

CONTENTS

Caroline Herschel
(Courtesy of Granger Collection)

ONE

The Cinderella of the Family

When Caroline Herschel was ninety-years-old she wrote about the first time she saw a comet. She was a young girl, still living at home, when her father had sneaked her out of the house to watch the comet cut across the night sky. They had to sneak because Caroline's mother, Anna, thought a girl's place was in the kitchen and doing housework. She did not want her daughter to be educated, or to have her head filled with the sort of wild ideas that could come from studying the stars, planets and comets. But Isaac Herschel knew Caroline was intelligent and curious. He liked to teach her about science and music and mathematics. Nearly a century later Caroline remembered that they slipped into "the street, to make me acquainted with some of the most beautiful constellations, after we had been gazing at a Comet." It was an exciting night for both father and daughter, but neither could have known, or even dreamed, that she would one day earn fame as an astronomer.

Comets

Chinese astronomers were the first to record the appearance of a comet, more than 4,000 years ago. The early Chinese called comets broom stars, and thought the gods used them to sweep away evil. By contrast, for many years, Europeans considered comets harbingers of doom. Today, we know that comets are made up mostly of rock, dust, and frozen chemicals (mostly water, and some carbon dioxide, and methane) mixed with some rocks and minerals.

Astronomers think billions of comets might be orbiting the sun at ranges up to three trillion miles (5.5 trillion km) away from it, in a spherical shell called the Oort cloud. There is also another doughnut-shaped swarm of comets closer to Earth, called the Kuiper belt located just beyond Neptune. On rare occasions, the gravity of a passing star pulls a comet out of the Oort cloud and flings it on a path towards the sun. Comets from the Kuiper belt can be nudged towards the sun by the gravity of a planet. Every year, a few hundred comets may be thrown towards the sun—but most are too small to be seen from Earth with the naked eye.

To travel from the Oort cloud to the inner solar system takes millions of years. Some comets loop just once towards the Sun and then move away, never to return. Some are captured in elliptical orbits around the sun. These comets are periodically visible from Earth. Once diverted, as the comet makes its way into the realm of the planets, it could pass close enough to one of the giant planets, such as Jupiter, to have its orbit changed dramatically. It could even be captured into orbit around the planet, or even more dramatically, as happened with the Comet Shoemaker-Levy 9 and Jupiter in 1994, crash into the planet. As a comet approaches the sun, it might also develop jets that change its orbit.

In the outer solar system, these objects appear as icy asteroids —irregularly shaped masses of dirty ice—but if their orbits are changed such that they move near the sun, their appearance changes dramatically.

The icy, hard part of a comet is called the nucleus. When a comet nears the sun, frozen chemicals turn into gas, and dust is also released. This forms a halo around the nucleus, called a coma. When a comet passes near the sun, its glowing coma may be visible from Earth, if it is large and bright enough. Though the coma usually brightens gradually as the comet approaches the sun, it can also light up quickly due to outbursts of activity from the nucleus—making it seem like a comet suddenly appeared in the sky from nowhere. The solar wind (charged particles that stream out of the sun) blows gas and dust away from the coma—forming one or more tails. Tails may stream in front or behind a comet, but always point away from the sun. Tails can be as much as one-hundred-million-kilometers long, and the coma can be up to one-million-kilometers wide, and the nucleus about ten kilometers in diameter.

Comets can travel around the sun at about one hundred sixty thousand kilometers per hour, making them some of the fastest objects in the solar system. The closer a comet comes to the Sun, the faster it moves. The point in a comet's orbit at which it is closest to the sun is known as perihelion; the point at which a comet is farthest away from the sun is called aphelion. Comet orbits are usually very elongated compared to the orbits of planets, so the distance between perihelion and aphelion for a comet may be anywhere from a few times to many thousands of times the diameter of Earth's orbit.

Comets that complete their orbits within two hundred years are called periodic comets. Those that take longer are called nonperiodic comets, because they have only been observed in close passage to the sun one time. About two hundred short-period comets are currently known. Once a comet has been discovered, measuring its position is very important. Astronomers need at least three different measurements to calculate a comet's orbit.

Comets tend to shrink in size as they make several trips around the sun because as its ice is vaporized, some of the gas and dust is pushed away by the solar wind. The rocky part of a comet may also be broken up into tiny fragments because it is subjected repeatedly to the pulling power of the sun. Comets may end their lives as a small rocky lump or as several lumps held together by gravity, or even crash into the sun or collide with a planet and be completely destroyed.

Caroline Lucretia Herschel, who was nicknamed Lina, was born in the German city of Hanover, on March 16, 1750. She was the eighth child, and fourth daughter, of Isaac and Anna Herschel. Her family had lived near Hanover ever since Caroline's great-grandfather, Abraham Herschel, had been expelled from another city because he was a Protestant and the area was predominantly Catholic.

In their new home Abraham's sons became farmers and gardeners. Isaac Herschel, Abraham's grandson and Caroline's father, learned how to be a gardener, but did not enjoy it. His love was music. He taught himself to play an old violin, and when he grew older, used the money he earned from gardening to pay for oboe lessons. Talented and determined, he practiced tirelessly. His family, certain that working the land was a secure profession because everyone needed food, discouraged him. But Isaac was determined, and with a characteristic perseverance that he would pass on to Caroline, continued practicing until he was good enough to become a military musician.

A year after leaving the farm, in 1732, Isaac married Anna Isle Moritzen, who soon gave birth to their first child, a daughter. Isaac and Anna ultimately had ten children together, six of whom survived to adulthood.

Although Isaac's pay was modest, he rose rapidly in rank until he was promoted to bandmaster. Life was good for a while. The country was at peace, Isaac's duties were not demanding, and he was able to give music lessons to supplement his income and still have time to spend with his family.

Unlike many members of the working class, Isaac instilled in his children a love for learning as well as a deep appreciation of music. He taught them his strong work ethic and told them "it was only in the thought of having done" their duty that they "could expect to find contentment." This sense of duty both helped and hindered Caroline when she became an adult.

Isaac encouraged his sons' intellectual pursuits and enhanced their education as much as he could. He wanted them to become eminent musicians and paid the most attention to their musical training, but he also loved astronomy and taught them all he knew about the heavens.

Isaac also encouraged the education of his daughters and sent them to school. Unfortunately, the teachers concentrated on the boys. Girls as a rule were given very rudimentary educations. They were instructed about their religious duties and taught to read and write, but were not usually exposed to arithmetic. Boys, however, were taught mathematics, and the more talented male pupils learned Latin.

Caroline's mother, Anna, was one of those opposed to the schooling of women. She was illiterate herself and disliked the thought of her daughters learning anything more than knitting and housework. Though she agreed to let Caroline attend school as long as Isaac was in Hanover, she kept Caroline so busy with household chores after classes that it interfered with her schoolwork.

Fredrick II ruled Prussia from 1740 to 1786.

In 1740, Fredrick II became the ruler of Prussia. Soon Europe was plunged into war. Isaac, now the father of five, was frequently forced to leave with his regiment, the Hanoverian Foot Guards. Although bandsmen were not expected to take part in the fighting and were allowed to take shelter as soon as the shooting started, they suffered from the same harsh living conditions. Isaac was sometimes forced to hide in cold, waterlogged ditches as the fighting raged, and before long his health began to suffer.

By the time Caroline was born, the eldest Herschel child, seventeen-year-old Sophia, had moved out of the Herschel household and worked as a maid for another family. Two other daughters born before Caroline had died as infants, which meant Caroline was the only girl at home. In Anna's eyes, Caroline was her last chance to have free domestic help. As soon as Caroline was old enough, her mother set about turning her into the family servant.

Fate seemed to be on Anna's side when it came to keeping Caroline servile. When she was three years old, Caroline was stricken with the same smallpox that killed one of her brothers. She survived, but her face was scarred and her left eye was misshapen. Anna knew a disfigured daughter would be difficult to marry off and did everything in her power to prevent Caroline from learning "more than was necessary for being useful in the family." Another advantage Anna had was that Isaac was often away from home.

Caroline's domestic training began early. She began knitting stockings for her brothers at such an early age "the first pair . . . touched the floor" while she "stood upright finishing the front!"

Anna showed Caroline few signs of affection. This, along with her mother's conviction that it was Caroline's duty to serve her brothers, as well as the ravages left by smallpox, had a powerful shaping influence on Caroline. All her life, she sacrificed her own desires so she could give her brother William her undivided attention. After William's death, when Caroline had the financial means to live a comfortable and more independent life, she instead moved into the household of her youngest brother, Dietrich, offering to "give herself, with all she was worth, to him and his family."

Dietrich, who was born in 1755, when Caroline was five, was dear to Caroline. But her favorite brother was William, who was nearly twelve years her senior. William had followed in his father's footsteps and became an oboist in the Hanoverian Foot Guards. He was often away at war, but when home he was attentive toward Caroline. Caroline was not close to Alexander, her next eldest sibling, or to Sophia. Jacob, her oldest brother, treated her with contempt, cuffing her whenever he thought it was necessary.

Sir Isaac Newton

When Caroline was six, an incident occurred that endeared William to her even more deeply. The Hanoverian Guard was returning after a long campaign. Anna was busy cooking an evening meal to welcome them back and sent Caroline to greet them. Caroline stayed out in the cold for hours, but could not spot her father or brothers. When she came home, she found them enjoying their meal. No one had noticed her absence or entry—except William, who "threw down his Knife and Fork" and "came running and crouched down" to hug her. So rare were signs of affection in Caroline's young life that this simple gesture made her forget the ill-treatment she often endured.

Isaac was kind to his daughter when he was home. He encouraged her as much as he could and taught her to play the violin. But her mother made sure that she had very little time to practice.

When Caroline's father and elder brothers were home, they often discussed science and philosophy late into the night. She would lie awake listening to them as they argued over the ideas of mathematicians and scientists such as Rene Descartes, Sir Isaac Newton, and Leonhard Euler.

In the spring of 1757, war-clouds loomed again, and the Hanoverian Guards were summoned to battle. Hanover was now officially under the English crown because the English King, George II, was also the elector, or ruler, of Hanover. Caroline watched as her father and brothers marched away, to the sound of beating drums and roaring troops. That summer, when the Hanoverians and their allies fought the French at the Battle of Hastenbeck, William and the other musicians sought shelter as cannons and rifles fired incessantly.

Caroline would lie in bed at night and listen to her father and brothers discuss leading scientists and mathematicians of the era, such as Leonhard Euler.

By autumn, the Hanoverian army had been defeated by the French, and its troops were fleeing and scattering in all directions. Caroline later recollected that she had:

In 1757 Hanover became a part of England. George II was the king of England and the ruler of Hanover.

Only by chance caught a parting glimpse of my brother [William] as I was sitting on the entrance of our street-door; when he glided like a shadow along, wrapt [wrapped] in a great coat followed by my Mother with a parcel containing his accoutrements, and after having succeeded to get unnoticed beyond the last centinel [sentinel] at Herrenhausen my brother changed dress—my mother returned to us alife [alive], but heartbroken with the performance of the painful task, with which on account of the required secrecy no one else could have been trusted.

As a seven-year-old, Caroline did not quite understand the significance of what was happening or why William was so strangely dressed. Later, she learned that he had been in disguise. This had been the first step in his escape to England

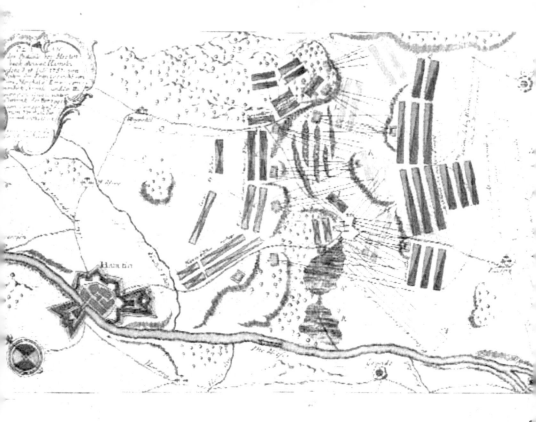

This drawing depicts the Battle of Hastenbeck, where in 1757 the French defeated the allied forces of Hanover, Hesse, and Brunswick.

where he and his parents hoped he could lead a better life and would not have to go to war. William reached England in safety and eventually decided to make it his home. Caroline seldom saw him for many years.

In the aftermath of the French victory, more than sixteen enemy soldiers took up residence in the modest apartment where Caroline lived with her mother, sister, Alexander, and Dietrich. Isaac and the rest of his defeated regiment were unable to return to Hanover because it was now an occupied territory. A year later, the French army marched away to battle again, abandoning Hanover after first destroying the food supply and cutting down nearby forests. The inhabitants were left hungry and cold, and soon the dreaded disease typhus started to ravage the town. These years were especially bleak for Caroline because, as she wrote later, the absence of William and Isaac meant "there was no one who cared anything about me." Caroline later destroyed all her written accounts of the difficult years of the French occupation and its aftermath.

In 1759, the Hanoverian army finally won a decisive victory over the French, and Hanover was not threatened again during Caroline's childhood. Isaac returned home and applied for a discharge from the Guards. Caroline's life improved. Her loving father was at her side once more, eager to try to give her "something like a polished education."

Then, in 1761, ten-year-old Caroline contracted typhus and almost died. Typhus weakened her so severely that for several months she had to climb stairs on her hands and feet. When she finally recovered, she found that the disease had stunted her growth. She never grew taller than four feet and three inches, and remained frail all her life.

Isaac became certain his daughter would never marry because of her pockmarked face and tiny stature. He did not try to hide this opinion from Caroline because he felt it was his duty to repeatedly warn her "against all thoughts of marr[y]ing saying as I was neither handsom[e] nor rich, it was not likely that any one would make me an offer."

Army life had taken a toll on Isaac's health, which was failing rapidly. In 1764, after suffering a paralytic seizure, he struggled on, giving music lessons and copying out music while Caroline read to him. William visited Isaac and assured him that he was doing well in England. William's arrival coincided with a solar eclipse. Because looking directly into the sun is dangerous, Isaac arranged for his family to observe the eclipse reflected "in a Tub of water . . . in the courtyard." During William's stay, Anna kept Caroline so busy "doing the drudgery of the scullery" that she had little time to spend with her favorite brother.

Caroline did not meekly accept a life of domestic servitude. Isaac's illness made her take further stock of her situation. She wrote: "I could not help feeling troubled sometimes about my future destiny . . . for I could not bear the idea of being turned into an Abigail or Housmaid [housemaid]." She decided to take steps to learn embroidery, hoping that with this skill, and "my little Notion of Music, I might obtain a

This map depicts the "Dark Shadows" of the solar eclipse that occurred while William was home for a visit.

place as a Governess in some Family where the want of a knowledge of French would be no objection."

Anna had purposely prevented Caroline from learning French and fancy needlework to close off her chances of becoming a governess. Secretly, Caroline arranged for lessons from a sympathetic neighbor. At dawn, the neighbor would cough, a signal that she was ready. Caroline would make a quick trip to her house for an early morning embroidery session. Unfortunately, this kindly friend soon died of tuberculosis.

Isaac died when Caroline was seventeen. She was devastated and completely at the mercy of her mother. Jacob, the oldest living son, was the new head of the family. He was a vain court musician, selfish, cruel, and extravagant. Caroline suffered "many a w[h]ipping" from him whenever she accidentally annoyed or irritated him.

Caroline could have grown bitter or given way to despair. Instead, she looked for a way to make herself financially independent. Resolving to learn a trade, she pleaded with and pestered Anna and Jacob to allow her to take lessons in dressmaking and millinery, saying she could use her new skills to serve the family. Anna grudgingly agreed, and managed to persuade a Mme Kuster, who ran a dressmaking class, to take Caroline on as a student.

Caroline was pleasantly surprised by her reception at the class. "Among the group were several young Ladies of genteel families and as I came there on rather reduced terms I expected no other, but that I would be kept back with doing nothing but the plain work of the business; but contrary to my fears, I gained in the school mistress a valuable friend which however made the reluctance with which I was obliged to relinquish her society after so short an acquaintance the greater."

Although Caroline had cajoled her family into letting her learn a few skills, she was worried that she would always be held back by her lack of a solid education. Then the possibility of a bright future arrived in the form of a letter from William. He had secured a prestigious job as an organist in the fashionable English town of Bath. Alexander, who had joined him there, had told him of their little sister's plight. In the letter, William wrote that Alexander had mentioned Caroline's musical talent. He suggested that Caroline come to England, where she could "become a useful singer for his winter Concerts and Oratorios," and asked Jacob to help prepare her by giving her singing lessons.

"This at first seemed agreeable to all parties but by the time I had set my heart upon this change in my situation Jacob began to turn the whole scheem [scheme] into ridicule." He refused to teach Caroline to sing. Ignoring Jacob's jibes and taunts, Caroline started to sing with a gag in her mouth, because she knew that professional singers practiced this way. With a perseverance that she would later use to master mathematics, she began "taking in the first place every opportunity when all were from home, to imitate with a gag between my teeth, the solo parts of Concertos . . . such as I had heard them play on the Violin; in consequence I had gained a tolerable execution before I knew how to sing."

Months of uncertainty passed as Jacob, Anna, and William argued about Caroline's future. She herself was sometimes in two minds. A part of her wanted to leave and learn new things; another part of her was frightened that she might disappoint William and fail in England. She did not even speak English.

Bath, England, circa late 1700s

Caroline sometimes felt so guilty about wanting to be free from the clutches of her mother and Jacob that she "knitted as many Cotton stockings as were to last two years at least" and also made ruffles that would be a gift for William's kind gesture if she stayed—and for Jacob if she went.

William came back to Hanover on August 2, 1772, determined to convince Anna and Jacob to let Caroline come with him. Jacob was not home. He had been called away to play for the court orchestra. Although it was Caroline's dearest wish to go to England and become a financially independent musician, she was uncomfortable about leaving without Jacob's permission.

As William's visit drew to a close, it became clear that Jacob would not return before he left. William saw that Anna used Caroline as a maid, and came up with a way to overcome his mother's objections. He promised to pay Anna enough to employ a servant to take over Caroline's duties. As soon as Anna felt assured that she would be financially compensated, she stopped resisting Caroline leaving.

All that remained was for Caroline to grasp the opportunity that lay before her. After an internal struggle, she decided to defy tradition and leave without her eldest brother's permission. It was sad to leave her mother, for whom, despite everything, she retained some affection.

Caroline later described her first twenty-one years of life as those that "had been sacrificed to the service of my family under the utmost self-privation without the least prospect or hope of future reward," and referred to herself as "the Cinderella of the family." On August 16, 1772, this Cinderella climbed into a carriage that carried her away from the drudgery of her household hearth and began the journey that took her eventually to the stars and the comets and helped lay the groundwork for new explorations of the structure of the universe.

TWO

Escape to England

The journey to England was far from comfortable, even for a seasoned traveler such as William, let alone Caroline, who had seldom journeyed outside Hanover. They spent six nights in a postwagen—an open stage coach—heading west to Holland, where they would board a boat to England.

William was delighted by the unbroken panorama of sky he could observe as the coach raced along during the night. They stopped only for meals and to change horses. He took time to re-acquaint Caroline with the constellations their father had introduced to her but she was less fascinated with the heavens than William and tried to sleep through the cool nights.

They finally crossed into Holland, where the fierce Dutch wind blew unobstructed over the flat landscape. To Caroline's dismay, her hat flew away as they crossed over a canal.

At the Dutch port of Helvoetsluys they embarked in a decrepit packet boat. A storm gathered and the waves surged and battered the boat as they crossed to England. By the

Caroline and William traveled from Holland to England aboard a packet boat similar to the one pictured here.

time they reached the port city of Yarmouth, the vessel's main mast and second mast had broken off. They completed the final yards of the voyage in a small open boat and were unceremoniously "thrown like balls by two sailors" onto the English coast.

After walking to a nearby house, where she enjoyed steaming tea and her first taste of fine English bread, Caroline and William got into a cart with other passengers, which dropped them off on a road where London-bound stagecoaches stopped for passengers. They were picked up by a coach drawn by a nervous, poorly trained horse and during the journey the horse bolted. Caroline, William, and another passenger threw themselves out. Fortuitously, Caroline landed in a dry ditch, and neither she nor the other passengers sustained serious injury. A gentleman and his servant came to their rescue eventually and escorted the coach the rest of the way. They arrived in London at midday, ten days after beginning their journey.

Caroline visited St. Paul's Cathedral during her brief stay in London.

William left Caroline at an inn to attend to some business. He returned in the evening and took her on a tour of the city. London was a massive city, many times the size of Hanover. The windows of the shops were lit, and Caroline noticed that her brother studied the lenses on display in any optician's shop that they passed. He explained that he had an interest in all types of lenses—particularly those that could be fitted into telescopes to magnify the stars and the planets. In order to learn astronomy, he said, one had to be able to see far away objects.

They also saw many of the city's more famous sights. Caroline was impressed by St. Paul's Cathedral, and the fine old Bank of England.

They left London the next evening by night coach for William's home in Bath, arriving the following afternoon. Caroline had hoped to be welcomed by their brother Alexander, but he was away. Instead, they were greeted by the Bulman family that occupied the ground floor of William's home.

Exhausted after the voyage of nearly two weeks, during which she had only slept two nights in a bed, Caroline retired to her room on the topmost floor that she would share with Alexander. She did not awake until the next afternoon.

The next day, William "began imediately [immediately] giving me a lesson in English and Arithetic [arithmetic], and shewing [showing] me the way of booking and keeping account of Cash received and laid out. The remainder of the forenoon was chiefly spent at the Harpsichord; shewing [showing] me the way how to practice singing with a Gag in my mouth." William prepared lessons for Caroline every morning, over breakfast. Unfortunately, William was not the best teacher. According to Caroline he did not grasp how poorly educated she was. "[W]e began generally with what we should have ended; he supposing I knew all that went before." She was forced to learn largely on her own.

William gave his sister a purse of English money for the household expenditures. She had to enter her purchases into a book. She went to the market on her own, although she was still learning English and had not mastered her multiplication tables—an accomplishment that eluded her all her life—and hated the bustle and din created by the crowds. Alexander faithfully followed his sister at a distance, keeping his presence a secret, though ready to come to her aid if needed. In the beginning, shopping was such a fearful undertaking she often "brought home whatever in my fright I could pick up." Eventually, she developed the confidence, and the skills, to shop on her own—but she was never very good it. Once, on a trip to London, she bought two horses, and then "had the mortification to hear . . . that they were both blind!"

Caroline was responsible for managing the servants, most of whom were very lazy. She expected them to work as hard as she had for her mother back in Hanover. She tried to pass off her strict instructions as William's, rather than her own, but soon after her arrival, the housemaid received her orders "with so much ill will . . . she gave warning and left us at Christmas." Caroline did not prove much better at supervising servants than at selecting horses, and the Herschels were subjected to a succession of "Pickpockets and Streetwalkers." At times, they were without servants altogether, and her relationships with the servants were never very good. This she blamed entirely on them: "very seldom I have been so fortunate as to meet with gratitude or good will in any [servants] I have had to deal with, and many times I could not help thinking but that it was owing to a natural antipathy the lower class of the English have against foreigners." Caroline never admitted to being even partially responsible for the failure of this, or any other relationship.

When her time was not taken up with household duties, Caroline did all she could to foster her career as a singer. William taught and trained her when he was able to spare the time. Until October, when the season began at Bath, she had a fair share of William's attention, and he "had leisure to try my abilities of making a useful singer for his concerts and oratorios of me, and being very well satisfyed [satisfied] with my voice I had 2 or 3 lessons every day." When they were not engaged with lessons, they relaxed by talking about astronomy.

As the winter music season approached, however, both of her brothers became busy. They were occupied with concerts and musical engagements and giving lessons in cello, oboe, violin, harpsichord and voice, in William's drawing room. She and William met almost exclusively at breakfast, and William coached Caroline as she sang and he ate, after which she practiced for six or more hours on the harpsichord.

As the season drew closer, she soon "began to fear when I saw one care after another devolving upon me that the hours required for practice would soon be much abridged, and that my Brother would not have much time to spare for giving me many lessons, for the Town soon began to fill and except at meal times he was seldom at home."

As the winter season progressed, Caroline saw very little of her brothers. In a rare criticism of William, she wrote unhappily: "I still was and remained almost throughout my long life without a Friend to whom I could have turned for comfort or advice when I was surrounded by trouble and difficulties. This was perhaps in consequence of my very dependent situation, for I never was allowed to form any acquaintance with any other but such as were agreeable to my . . . brother." Although Caroline adored and admired William, she never considered him a confidante, and in fact made "the resolution of never opening my lips to my dear brother William about worldly or serious concerns." Alexander spent time with her, but Caroline wrote that it "did me no good, for he never was of a cheerful disposition, but always looking on the dark side of every thing, and I was much disheartened."

Caroline's first winter in England was, overall, not a happy one. She was a foreigner, far from the only home she had known for the first time in her life. Despite the kindness of William and Alexander, and the beauty of the city of Bath, she struggled against homesickness. Caroline, who had spoken German all her life, also knew only a few English words, making it uncomfortable for her among strangers. To make matters worse, Caroline pronounced the Bulmans' daughter, who was only a few years younger than her, "very little better than an idiot" and she referred to William's servant as a "hot-headed old Welsh woman."

When spring arrived, Caroline hoped to receive more instruction and attention from William. However, William often came home exhausted by his work, and Caroline helped him retire "with a Bason [Basin] of Milk or Tea—and Smith's Harmonics or Optics, Ferguson[']s Astronomy, and so went to sleep buried under his favourit[e] Authors." In the mornings he usually gave his family "an astronomical Lecture of which Alexander was generally a partaker."

Astronomy was fast becoming William's obsession. Merely reading books was not enough. William wanted to become a watcher of the heavens, for he "was not contented with knowing what former observers had seen." He spent the summer of 1773 buying lenses and trying to make telescope tubes. Caroline was soon assisting him, "with making the tube of pasteboard against the glasses arrived from London, for at that time no Optician had settled at Bath." Caroline's first task as an astronomer failed, because pasteboard was not strong enough to be a tube for a durable telescope.

Prefabricated telescope supplies were not the quality William wanted. This was partly because during William's time most European astronomers were preoccupied with the question of whether gravity was universal and spent most of their time trying to verify that the planets behaved as the law of gravitation Newton had set forth predicted. To see if planets moved as expected, it was necessary to accurately measure their positions—where they were, and when. It didn't matter how much a telescope magnified them, because the focus was on planetary motion— not what planets looked like or were made of. Serious astronomers of the time also restricted their investigations to the solar system, and spend little time studying the stars or the rest of the universe.

As a self-taught astronomer, William had not adopted most of the prevailing scientific prejudices of his day. Unwittingly, he struck out on a path that would change the focus of astronomy. He decided to study distant objects and to study the nature of heavenly bodies, in addition to their movements.

Objects, such as stars, that were in the deep sky were much fainter than objects within the solar system, such as planets. To study the characteristics of faraway objects,

William wanted large telescopes—instruments that could collect more light than the smaller objects then in use.

There were two main types of optical telescopes: refracting telescopes and reflecting telescopes. Refracting telescopes have lenses and no mirrors and the light that falls on a lens at the top of the telescope tube is focused near the bottom. The observer looks at the focused image through an eyepiece. These were the most commonly used telescopes when William turned to astronomy.

In reflecting telescopes, light passes down to the bottom of the tube. At the bottom a mirror reflects the light. This allowed the image to be focused near the top where it was examined by an eyepiece. Reflecting telescopes could offer a clearer focus and higher magnification than refracting telescopes. The lenses of refracting telescopes acted like prisms and produced images with colored edges; reflectors did not do this.

It was clear to William that refracting telescopes could not meet his needs, both because of the color separation and

Telescope Properties

Astronomical telescopes are designed to make distant objects easier to see and study. To most people, this means that they magnify views of distant objects. This is certainly true, but for most astronomers, magnification may be the least important of a trio of telescope properties.

Most professional astronomical telescopes are very large. William Herschel started a revolution in astronomy by building what was, for his time, exceedingly large mirrors for his reflecting telescopes. The trend continues to this day, as the largest telescope in the world, at the Keck Observatory atop Mauna Kea in Hawaii, uses a ten-meter wide complex mirror atop a mountain in Hawaii. In fact, there are two ten-meter Keck telescopes side by side on the mountain.

Large mirrors provide two advantages. First, their large surface areas gather and focus more light than a smaller mirror can collect. More light down the tube means that a brighter image is formed, so large telescopes are well-suited for viewing faint objects. One reason so few nebulae had been discovered before William and Caroline began sweeping the heavens is because the instruments used by early astronomers were so small. Most nebulae are very faint, so it took a larger instrument to coax them out of the darkness.

The second advantage is resolving power (or resolution), which is the ability of the instrument to distinguish fine detail in an image. Larger telescopes provide better resolution than smaller ones. One area where high resolution is necessary is the study of close double stars. With a small telescope, the light of the two stars seems to smear together, making it hard to distinguish them as a pair, but a larger high resolution instrument shows them as a cleanly separated pair.

Many people confuse magnification with resolution. Magnification is simply how much larger an object appears through the telescope, but a magnified view of a low-resolution image is nothing but a large image of a blurry object. So while magnification sounds impressive, it is resolution and light gathering power that are the real measures of the ability of a telescope to advance our knowledge.

because glass manufacturers were unable to make lenses large enough to collect enough light for him to see faint faraway objects. He needed to use high-quality reflecting telescopes and decided to design and build his own instruments.

In September 1773, William bought secondhand tools for making reflectors. William expected Caroline and Alexander to help him in his new pursuit. Caroline found herself "much hindered in my practice by my help being continually wanted in the executing of the various contrivances." Her initial reaction to astronomy was irritation, rather than interest.

To manufacture reflecting telescopes, they had to make lenses and polish and mount their mirrors. Alexander, who had a mechanical bent of mind, assisted enthusiastically in constructing telescopes and manufacturing telescope parts. Caroline, on the other hand, was less than pleased to see the musical instruments and fine furnishings moved to make place for telescope tools: "to my sorrow I saw almost every room turned into a workshop. A Cabinet Maker making a Tube and stands of all descriptions in a handsome furnished drawing-room. Alex putting up a huge turning machine . . . in a bedroom for turning patterns, grinding glasses & turning eye pieces."

The Herschels learned that there were two important types of reflecting telescopes: Newtonian and Gregorian. The Newtonian reflector was named after Sir Isaac Newton, who invented it. In Newtonian telescopes, the image formed by the large mirror is reflected at an angle from a small, flat mirror (called the secondary mirror) to the side of the tube. The observer looks at the image through an eyepiece located at the side. In the Gregorian telescope, invented by Scottish astronomer and mathematician James Gregory a few years before Newton built his first reflector, the image made by the large mirror is reflected back by a curved, concave mirror, through a small hole in the center of the main mirror. The eyepiece is fitted into this small hole in the large mirror,

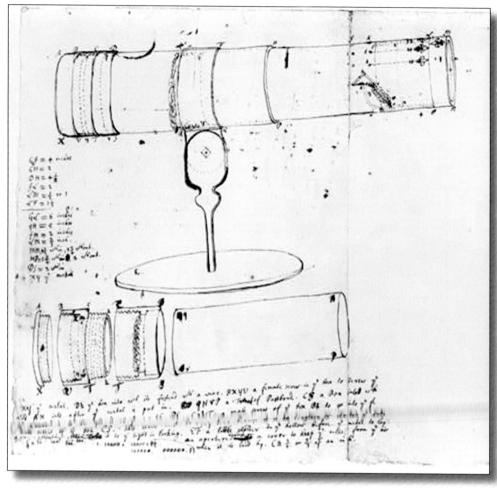

A sketch of the Newtonian model reflecting telescope

allowing the observer to look straightforward, from below, rather than from the side. There were other variations of reflector design, but William focused on learning to construct Newtonian and Gregorians.

Despite his growing skill as an astronomer and telescope maker, William could not afford to pursue it full time. Instead, he relied on his still lucrative music career to support himself and his family. However, when the music season ended, William focused all his attention on astronomy. For a few years, the siblings followed a routine set by William, in which music (their profession) was the main winter occupation, and astronomy (William's passion) the main summer occupation.

Despite the demands on William's time, he was concerned about providing for Caroline's social as well as material needs. William wanted Caroline to settle into life, to enjoy

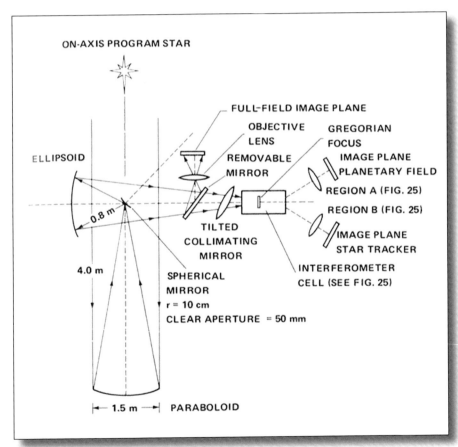

A detailed diagram of the pieces of a Gregorian telescope

One of the only images of Caroline Herschel when she was young.

singing, and to learn to sing well. He tried to build up her confidence and persuaded his shy little sister to sing whenever they had a party. He also asked her to take part in the rehearsals and services of his choir, held at the fashionable Octagon chapel.

Anxious that lonely Caroline make friends with ladies of refinement, William introduced her to his female acquaintances. But Caroline described them as "the wicked pilfering wretches by which my brother was surrounded," and declared that she was not much edified by their company. She disliked the idea that she had to pay other women compliments and was disdainful of Bath society, where female social life centered on flower shows and fancy balls.

William engaged Anne Fleming, Bath's most reputed dancing teacher, to turn his taciturn little sister into a sophisticated and polished woman. She spent time twice a week with Caroline "to drill me for a gentlewoman (God knows how she succeeded)."

Succeed she did. Caroline grew friendlier and more polite, however reluctant she might have been about the process. She was changing: growing into the dignified lady who would one day entertain royalty without the least sense of trepidation.

THREE

Stars and Song

By the spring of 1774, astronomy had become so important to William that the Herschels' home was not large enough and they moved to a new home. The new location had room for workshops, a place on the roof for observing, a plot that allowed an uninterrupted view of the sky, and space enough for the construction of a giant telescope. They had finished their first satisfactory telescope before the Bath music season of 1774 began. It was a five-and-a-half-foot long scope with a Gregorian design that William used to study the stars, entering his first nightly observation into his astronomical journal that year.

William and Alexander continued to build telescopes, sometimes, much to Caroline's dismay, in their best clothes: "every leisure moment was eagerly snatched at for resuming some work which was in progress, without taking time for changing dress, and many a lace ruffle . . . was torn or bespattered by molten pitch." One winter Sunday in 1775, the brothers stole out to work after church, but soon William

Drawing of a Herschel telescope

"was soon brought fainting back by Alex with the loss of a nail of one of his fingers."

Their second major telescope project was the construction of a seven-foot-long Newtonian with a magnification power of 222. In May 1776, William observed Saturn with this new reflector. Next, they designed a giant telescope, an instrument they later called the "small 20 ft." This reflector brought its image to focus twenty feet away from its mirror, near the top of the tube. The mirror was twelve inches in diameter. It was not the easiest to use, as William found on observing Saturn in July 1776. The observer had to stand on a ladder, nearly twenty feet above the ground, in the dark, and peer through the eyepiece that was near the top. The tube was just slung

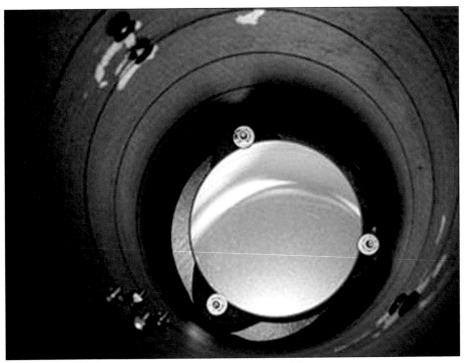

The mirror inside a Newtonian telescope

from a pole. The crude design of the scope indicated they still had a lot to learn about telescope design.

Slowly, William acquired considerable skill at making mirrors. The quality of the mirror and the eyepiece were critically important. Once William began to grind and polish a mirror, he refused to let go of it, because he realized that it was a craft that required a feel for the mirror's shape, which he feared would be lost if he took his hands away even for an instant. Making a mirror took many hours and it was left to Caroline to ensure that her brother ate and drank, even when he refused to use his own hands. She wrote:

> [My time] was so much taken up with copying Music and practicing, besides attendance on my Brother when polishing, that by way of keeping him alife [alive] I was even obliged to feed him by putting the Vitals by bitts [bits] into his mouth—this was once the case when at the finishing of a 7 feet mirror he had not left his hands from it for 16 hours together. . . . And generally I was obliged to read to him when at some work which required no thinking, and sometimes lending a hand, I became in time as useful a member of the workshop as a boy might be to his master in the first year of his apprenticeship.

Caroline did record progress in learning mathematics beyond the rudimentary. Her lessons included exercises in algebra, geometry, trigonometry, and spherical trigonometry. William's teaching skills had not improved, however, and the lessons were no more thorough than when they had started. Caroline was still forced to gather her knowledge whenever and however she could. She faithfully finished every exercise he gave her and asked him questions about mathematics and astronomy at breakfast, recording his answers in a book. Her book soon grew into "a miscellaneous jumble of elementary formulae, solutions of problems in trigonometry, rules for the use of tables of logarithms, for converting sidereal into solar time, and the like." She applied herself to her self-study of abstract concepts and, just as she had proved that an incomplete musical training was not an insurmountable impediment to becoming an excellent singer, she proved that she could master advanced mathematical skills despite the gaps in her early education.

William encouraged Caroline, often writing appreciative notes after he corrected her exercises. But he also occasionally punished, as though she were a child, making her sometimes give up her dessert if she did not correctly guess the angle of the piece of pudding she was helping herself to.

Caroline frequently lamented the loss of time for her musical practice. "I could not help feeling some uneasiness about my future prospects, for all the time my Brother could spare from his publick [public] business and attending on his Scholars was completely filled up with Optical and Mechanical works; and the fine nights with viewing the Heavens, so that I could not hope for receiving any lessons or direction in my practising."

Even with her lessons neglected, by 1777 Caroline was singing well enough for William to ask her to perform as one of the principal singers in a performance of Handel's "Judas Maccabaeus" that he was conducting. William gave her ten guineas to spend on a suitable dress. Her performance was a resounding success. The proprietor of the Bath Theatre called her as an ornament to the stage and the Marchioness

Royal Astronomer Nevil Maskelyne visited the Herschels in 1777.

of Lothian complimented her "for speaking my words like an English woman." Caroline's musical career was on the rise, although she was upset at being unable to devote as much time to music as she wished.

After this successful season astronomy invaded the home once more. Caroline resented the intrusion but was too subservient to do anything but bow to William's wishes.

The summer of 1777 was not peaceful, however. Word reached the Herschel siblings that their youngest brother, Dietrich, had run away from home. William left for continental Europe to try and track down Dietrich and take him back home. In the meantime, Dietrich arrived in London sick. Alexander went to London and brought Dietrich to Bath, where he remained for two years. Though Caroline doted on her baby brother, she found that his presence took away from time she wanted to spend on improving herself. Her brothers saw her first and foremost as their helper, and Caroline also defined herself as a housekeeper first.

In 1777 the Astronomer Royal, Nevil Maskelyne, paid his first visit to the Herschels. William and Maskelyne "engaged in a long conversation which . . . sounded like Quarrelling, and the first words my Brother said after he was gone was, 'That is a Deavil [devil] of a fellow!'" William meant that as praise; he had enjoyed their lively debate. Maskelyne became a trusted colleague, and within a few years, many other scholars and distinguished men, among them Dr. Blagden, the secretary of the Royal Society, Sir Harry Englefield, and Sir William Watson, befriended the Herschels.

During the next music season, Caroline took first soprano parts in "Samson," "Judas Maccabaeus" and other oratorios. She sang as prima donna at winter concerts at Bristol as well as Bath. At the Easter rendition of Handel's *Messiah*, conducted by William, Caroline's performance as first soloist was so stunning that she was invited to sing at a prestigious music festival in Birmingham. Her dream of a financially independent future as a singer was within reach.

Yet, poised for success and at the peak of her musical power, Caroline did not take the next step. After working

Caroline and William frequently performed in Bristol, England. *(Library of Congress)*

so hard for this opportunity, she turned it down, saying she "never intended to sing any where, but where my Brother was the Conductor." She could not conceive of striking out on her own without William, or leaving his side for any great length of time. The opportunity to do so never arose again.

Caroline had chosen to close the door on her musical career, her heart's desire. She sacrificed music of her own volition, but she did so with sadness. Though the decision was hers alone, she was conflicted and wrote resentfully about it, but did not blame her brother, herself, or her traditional upbringing.

Instead, she blamed William's servants, insinuating that it was they who claimed her time and brought about the decline in her musical abilities. She described this as the greatest grievance of her life: "I have been throughout annoyed and hindered in my endeavours at perfecting myself in any branch of knowledge by which I could hope to gain a creditable livelihood; on account of continual interruption in my practice by being obliged to keep order in a family on which I was myself a dependent."

That Caroline threw away the opportunity to accept the prestigious musical engagement was a testament to her devotion to William. In the end, this devotion shaped her destiny.

FOUR

Watcher of the Skies

In the spring of 1779, Caroline sang for the last time as a principal in oratorios. Meanwhile, William began to encourage her to become an astronomical observer in her own right and polished the mirrors of a Gregorian telescope he referred to as hers. Guests to the Herschel household reported that Caroline was stargazing on her own, even on cold nights.

William had familiarized himself with the night sky, and now it was Caroline's turn to master the map of the heavens. With constant practice, they both acquired dexterity at observational astronomy that helped them to make discoveries that were hidden from other observers.

Proper stargazing was, in some respects, an art that had to be learned. Anyone could look through a telescope; it took a skilled observer to identify the many heavenly bodies upon sight and distinguish them from an as yet unrecorded object. To make new discoveries, one required a trained eye, an in-depth practical knowledge of the stars, and an ability to recognize all known objects immediately.

In 1780, Caroline sang again, but her lack of practice was beginning to show. William was cutting back on his musical engagements and spending more time on astronomy. He had begun his second review of the heavens, his aim being to leave no spot in the sky unobserved.

On March 13, 1781, as William was looking at the Gemini constellation, he spotted a new object and discerned that it was not a point, like a star, but was a disc. He verified this by looking at it again under higher magnification—the object still appeared to have a disc, so he had to conclude that it was not a star.

William knew that this object had not been recorded previously and guessed, cautiously, that it was a comet and conveyed his discovery to the Royal Society. He noticed that the object moved from west to east, as most planets did, but was careful not to jump to any conclusions. Numerous comets had been discovered since the dawn of civilization, but finding a new planet was very rare. For centuries, few Western astronomers suspected the solar system contained planets that orbited the sun far beyond Saturn. It was almost unimaginable that the boundaries of the solar system extended any farther.

The Royal Society was formally incorporated in 1663, the first state-sponsored organization dedicated to the study of science.

Established astronomers began studying William's new object. William, too, was curious about what he had discovered, and he began to measure its movements. Everyone was intrigued about the nature of this object: was it merely a comet, or was it something else?

Months went by and observations of the object's position and movement accumulated. Soon there were enough data to mathematically compute its orbit. It was undeniable—amateur astronomer William Herschel had discovered a planet.

William was becoming a famous astronomer, and almost all of his—and Caroline's—time was devoted to the stargazing and the construction of telescopes. She wrote: "As soon as the public amusements were ended . . . my Brother applied himself to perfect his mirrors erecting in his garden a stand for his 20 ft Teles[cope]. . . . In short I saw and heard of nothing else . . . but about those things when my Brothers were together."

She performed many tasks in her assistance of her brother. She later wrote:

> Alexander was always very alert and assisting when anything new was going forward, but he wanted perseverance and never liked to confine himself at home for many hours together; and so it happened that my Brother was obliged to make trial of my abilities in copying for him catalogs, Tables . . . and sometimes whole papers which were lent him for his perusal . . . which kept me employed while my Brother was at the telescope at night; for when I found that a hand sometimes was wanted when any particular measures were to be made with the Lamp micrometer . . . and a fire to be kept in, and a dish of Coffe[e] necessary during a long nights watching; I undertook with pleasure what others might have thought a hardship.

William proposed that the Herschels build a thirty-foot telescope, with a mirror that was about three feet in diameter. This giant telescope would need a bigger mirror than any commercially manufactured in the world. No nearby foundry could undertake such an ambitious project, so the indefatigable William converted the basement of their home into a foundry.

Caroline was given primary responsibility for a long, odorous, and unenviable aspect of the making of the great telescope:

> The mirror was to be cast in a mould of lo[a]m prepared from horse dung of which an immense quantity was to be pounded in a morter [mortar] and sifted through a fine seaf [sieve]; it was an endless piece of work and served me for many an hours exercise and Alex frequent[t]ly took his turn at it, for we were all eager to do something towards the great undertaking. Even Sr Wm [Sir William Watson] would sometimes take the pestle from me when he found me in the workroom where he expected to find his friend.

Caroline did not take part in casting the mirror, as it was considered too dangerous an occupation for a female. She watched the proceedings with interest from a safe distance and saw the furnace begin to leak. Half a ton of molten metal ran into the fire and onto the floor of the washhouse. Caroline's "Brothers, and the Caster and his men were obliged to run out at opposite doors; for the stone flooring (which ought to have been taken up) flew about in all directions as high as the ceiling. My poor Brother fell exhausted by heat and exertion on a heap of brick-batts."

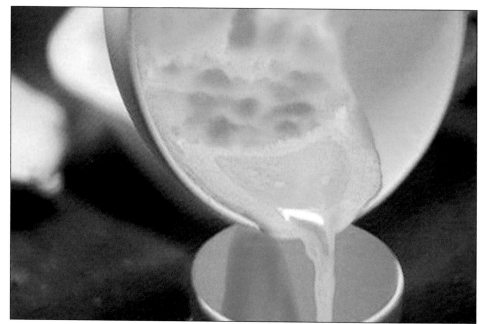

The Herschels had to work with dangerous molten metal in the construction of telescopes.

This setback did not deter them from trying again. William supervised the salvaging of the metal while Caroline pounded and sifted more horse dung. The second attempt produced what appeared to be a perfect casting until closer examination revealed that it had cracked upon cooling, and a third attempt was necessary.

Easter brought a temporary respite to the increasingly dangerous experiments. One morning during Passion Week, while they were preparing for a trip to Bristol where William was conducting a performance of *The Messiah*, their nephew George Greisbach, son of sister Sophia, arrived. He was now a member of the Court orchestra and had come to inform William that the king commanded William come to London with his now famous homemade telescope. William was understandably excited and distracted. Caroline assembled the music, and the performance proceeded, but it was clear that William was anxious to end the concert.

The Herschels gave their final performance in St. Margaret's Chapel, on Whit-Sunday in May, 1782. Caroline sang with William at the organ, performing one of his own compositions. Although Caroline had begun to use words such as eager to describe her participation in tasks associated with astronomy, she was unhappy as she "opened [her] mouth for the last time before the public."

The week after the curtain dropped on Caroline's final concert, William went to London to meet the king. He stayed at astronomer Sir William Watson's father's home. Watson also came to London to lend his support. They both hoped that the king would agree to give William a salary so that he could, at last, be paid to conduct astronomical research.

Caroline waited anxiously in Bath to

King George III offered William a salary to pursue astronomy.

hear the developments. The king was cautious and took his time. William wrote her three letters during his absence, but the letters revealed little.

Although William had discovered a planet, the king wanted to hear what Nevil Maskelyne, the fifth Royal Astronomer, and other leading astronomers thought of him and his telescope. He also wanted to verify that William was worthy of royal patronage. King George III interviewed William personally, and then asked William to set up his telescope at the Greenwich Royal Observatory. After William and his telescope successfully passed this trial, the king asked that the telescope be moved to Windsor Castle.

Luckily, William, unlike Caroline, was very tactful and had good interpersonal skills. The king enjoyed his visit and his chance to look through the telescope. Finally, more than two months after William's departure from Bath, the King made William an offer that would allow him to free himself of his musical obligations.

The offer was that William would receive a lifelong salary of two hundred pounds a year. In return, he would live near Windsor, pursue astronomical research, and allow the royal family and their visitors to look through his telescopes whenever they wished. William leaped at this opportunity to become a professional astronomer, although the salary was only about half of what he had been earning from music. Watson was shocked. "Never bought Monarch honour so cheap!" he commented, but William was content, especially as he was allowed to make additional money selling telescopes.

William named the planet he'd discovered Georgium Sidus [George's star] to honor his patron. He and Caroline would always refer to it as the Georgian planet, but the name did not stick. Jerôme Lalande of La Collége Royale [The Royal College] in France suggested that it should be called Herschel, after its discoverer but Bode, an astronomer from Berlin, suggested the name Uranus, after Urania, who was the Greek muse of astronomy. Uranus was also one of the old gods in Greek mythology, the grandfather of Zeus. This name was in keeping with the tradition of naming planets after gods and goddesses—though usually with the Latin names rather than Greek—and this was the one that survived.

Uranus

To the ancients there were five planets, or wandering stars: Mercury, Venus, Mars, Jupiter, and Saturn. The large orbs of the moon and sun held special status. When William Herschel first took to sweeping the heavens with his telescopes, there were six known planets, none of which could be claimed to have been discovered by any one person.

But in 1781 William Herschel discovered a faint orb moving through the depths of our solar system, well beyond Saturn. He dedicated his discovery to the King of England, by proposing the name *Georgium Sidus*, but such a nationalistic name didn't set well with astronomers in other countries, and eventually the name Uranus—after a Greek god of the sky, was chosen.

Uranus orbits the sun at a distance more than nineteen times the earth-sun distance, and takes eighty-four years to complete its circuit. All the while, it moves through space tipped on its side such that its poles are nearly in the plane of its orbit. For comparison, Earth's tilt is only twenty-three and a half degrees from vertical. With such an extreme tilt, polar regions of Uranus spend decades in sunlight, followed by decades in darkness. But the sunlight is feeble at such a great distance from the sun, and the temperature at the cloud tops of Uranus is around sixty degrees above absolute zero (roughly -350°F).

Uranus is one of the gas giant planets, with no solid surface and a composition dominated by hydrogen. Deep in its interior there is a liquid layer of water mixed with ammonia and salts, and a central core composed of the same kinds of rocky materials that Earth is made of, but under intense pressure. The clouds in the upper Uranian atmosphere are made mostly of methane ice crystals, which reflect blue light better than other colors, giving Uranus its ghostly blue tint.

Uranus is surrounded by a system of narrow rings that are made mostly of small chunks of dark ice, and twenty-seven moons have been found to orbit it.

After Herschel discovered Uranus, its motion was tracked carefully by astronomers worldwide. Eventually, small deviations of the new planet from its predicted path led to the suggestion that another large planet might lurk beyond Uranus. Application of the laws of gravity and orbital motion eventually led to the discovery of Neptune—just where the physical laws predicted it would be.

Impulsively, William found a house at Datchet, near Windsor, where the royal family had a residence, engaged a maid, and told Caroline that she and Alexander should move there at once. The house was an abandoned hunting lodge, ruined and neglected, with a leaky roof, an overgrown garden, and damp, dismal grounds. William had made his unwise choice because of the stables that he thought could be turned into a workshop for casting and polishing reflectors, and the laundry he had decided to convert into a library for his astronomical records. By August 1, 1782, the three Herschels were established at Datchet, to Caroline's displeasure.

Caroline had learned to enjoy her life at Bath. It had been sociable, with pupils arriving daily for their lessons, choirs for her to lead and train, and performances to practice for. She had made no close friends in Bath, but in Datchet life was even lonelier. The village was small and isolated.

In spite of this, Caroline continued to act as her brother's efficient housekeeper and work partner. Within three days of their arrival the household was running smoothly and William was able to resume his observations. The price of food at Datchet was high, but Caroline was thrifty and made sure William was not bothered by money matters. She also sat by him every night as he conducted his observations, ready to do whatever she could to help him.

Caroline soon discovered William had ambitious plans for her. It was now Caroline's sight, not her voice, that her brother needed, and he supplied her with "a Telescope adapted for sweeping consisting of a Tube with two glasses such as are commonly used in a finder." A finder was a smaller telescope attached to the larger one that pointed in the same direction. It was used for lining up the big telescope. The finder could see a wider section of the sky which made finding the right direction easier.

William suggested that Caroline search the night sky to look for comets and other undiscovered bodies. She was told to "sweep for Comets, and by my Journal No 1, I see that I began Augt 22, 1782, to write down and describe all remarkable appearances I saw in my Sweeps."

A sweep was a strip of sky. To observe every part of the night sky in a systematic manner, they divided the hemisphere of the night sky into imaginary narrow horizontal or vertical bands. For horizontal sweeping, all the stars within the band, or sweep, would be approximately the same angular distance from the zenith because sweeps were concentric circles, from around the zenith down to the horizon. Zenith is the point in the sky directly overhead the observer. As the earth rotates, this point moves through the stars, along a circle.

If sweeps were conducted vertically, the observer swept from zenith to the horizon (one sweep), then turned the telescope to point in a different direction and repeated the procedure. Vertical sweeps thus scanned the sky in bands that converged overhead, at the zenith. It was impossible to scan all of the visible heavens in one night, but a comet hunter's goal was to complete the procedure in as few nights as possible and then repeat it.

William also asked Caroline to look for double stars and nebulae. Double stars are two stars that appear so close together that to the naked eye they looked like one star. But with a powerful enough telescope, an observer could make out two distinct stars.

Nebulae (derived from the Latin word for hazy) are interstellar clouds of dust, gas, and particles. The particles of gas, dust, and matter in nebulae are too small to see but if there is a bright star within, or close, to a nebula its light illuminates the particles, making the nebula appear like a bright veil of mist.

Having learned from William about nebulae and different types of stars, Caroline began observing, but she did not

Double Stars

Our sun is a single stand-alone star, but many stars in our galaxy are members of multiple star systems. Before William and Caroline Herschel's time, most astronomers believed that if two stars appeared close together through

Double stars *(NASA)*

the telescope, it was because they were seen along the same line of sight. In other words, their apparent closeness was the result of a chance alignment of two stars that were far apart in space. Since many paired stars include one member that is brighter than the other, it was assumed by astronomers of Herschel's time that the brighter star was closer.

William wanted to use such alignments to measure the stellar parallax, the apparent shift of a relatively nearby star due to the changing position of Earth in its orbit. By carefully observing the separation between close doubles six months apart, he could tell if one star had appeared to shift, and if so, triangulate the distance to the star.

After cataloging hundreds of double stars he noticed during his sweeps between 1781-1784, William measured each of them closely and found no significant shifts. When he returned to this work in 1802, however, he found that some of the stars had changed their orientations since the 1780s. These were true double stars—stars that were gravitationally bound and orbiting each other. William presented this groundbreaking find to the Royal Society in 1803, and astronomers have been tracing the dances of these doubles around each other ever since. Modern catalogs document over 700,000 closely studied double stars.

Gravitationally related double stars are called binary stars. In such systems, the two stars revolve around a common center of mass. The time it takes them to complete their orbits depends on the distance they are apart. Some widely separated binary stars take thousands of years, while close pairs can take less than a day to wheel around each other.

Many of the most prominent stars in the sky are doubles. Sirius, the brightest nighttime star, has a faint compact companion in orbit around it. Polaris, the North Star, appears double through the telescope, but is actually a triple star system. The closest star to our solar system, Alpha Centauri, is also a multiple star, with a faint star circling a close double.

take to the pastime of stargazing immediately. It was often bitterly cold and she had to spend the long hours alone in the dark.

Despite her discomfort, she excelled at the task. In her first night of observation Caroline found a double star that William had studied earlier. By the end of her second night,

A nebula consists of gas, dust, and particles. Light from a nearby star illuminates the nebula.

she had discovered a previously listed nebula that William had never located.

Alexander decided it was time to part from his elder brother. Caroline and William tried to persuade him to take up residence at London, which was close to Datchet, and where he could continue his musical career, but he chose instead to return to Bath. Caroline described his parting as painful.

Despite the cold, lonely nights, by January of 1783 Caroline had started to enjoy watching the skies. It was essential that a comet-hunter identify a comet as soon as possible after it came into view, and to do so with confidence she would have to distinguish the new arrival from other fuzzy objects such as nebulae. William suggested that she familiarize herself with Messier's nebulae; every night, after sweeping, Caroline would spend time looking for a few Messier objects. Messier, an avid comet hunter, had meticulously cataloged many of the visible nebulae in order to not confuse these permanent features of the night sky with the comets that came and went.

A month later, on February 26, Caroline saw something that no one had ever recorded. It was a nebulous open cluster, in the constellation of Argo Navis, the ship, towards Canis Major. William confirmed her discovery and allowed Caroline to note proudly that Messier did not have it. (Modern scholars, however, have noted a discrepancy in Caroline's notes that leads them to believe that this object is actually Messier object number 93, discovered

Types of Nebulae

There are many specific types of nebulae in the universe. Clouds of dusty gas that shine by reflected light are called reflection nebulae. In some nebulae radiation from very hot young stars can make the gas glow. These nebulae actually emit their own light and are called emission nebulae. Other nebulae don't have stars within or near them to make them glow. These dusty nebulae may obscure our view of everything that lies beyond, and appear dark, like a dense fog. These are called dark nebulae.

Some nebulae are stellar nurseries. Star formation begins inside cold dark clouds of gas and dust. As areas within these clouds collapse under their own weight, they eventually produce concentrated masses of gas that are hot enough at their centers to produce their own energy—and stars are born. Nebulae in which this kind of process is going on may contain different areas of emission, reflection, and dark nebulae within the same complex structure.

At the end of the life of a small star, such as the sun, it undergoes expansions and contractions during which some of its hot, glowing gases are sloughed off to form a shell around the star. These shells are referred to as planetary nebulae, because they look like little disks through telescopes, resembling planets, although they are not planets. They received their name from William Herschel, who thought their round, often blue-green, appearance reminded him of the disk presented by Uranus, the planet he discovered in 1781,

Massive stars may explode at the end of their lives. Stellar explosions are called novae and supernovae. After a supernova explosion, the gases that were once part of the star may glow for thousands of years. These supernova remnants are also called nebulae.

As time goes on, the particles of matter that once formed the nebula around a dead star spread out. These particles may drift apart and remain dispersed for eons, but the diffuse particles can also contract again as a result of shock waves sent out by the death of a nearby star. If the shock waves push particles close enough together they can coalesce into nebulae that become new stellar nurseries, and stellar evolution can come full circle—the death of large stars leading to the birth of new ones.

The Hubble Space Telescope captured part of the Cone Emission Nebula.

by Messier in 1791.) A few minutes later, she saw another, very faint nebula near Canis Major—another that Messier had not recorded. Caroline had, on her own, made two discoveries in one night.

This was an amazing achievement, especially given that Caroline was just starting out as an astronomer. William recorded the nebulae and gave his sister all due credit. Yet her discovery did not make much of an impression in the astronomical world. Nebulae were not considered as interesting, or important, as comets. William's notes on his sister's discovery of nebulae were actually addenda to his papers.

In March of 1783, Caroline found more objects Messier had missed. She noted down the discovery of a cluster in Monoceros and an open cluster in Hydra. By the end of July, she had found two more objects; one an open cluster in Ophiuchus.

On August 27, 1783, Caroline made yet another important discovery. While observing the Messier object M 31 (the Andromeda Nebula), she discovered that it had a companion Messier had not included in his famous list (though his records show he had observed it in 1773, and it is now included in most publications of Messier's catalog as number 110, the last object on the list). In September, Caroline found another new object in the constellation Sculptor.

She continued making deep-sky discoveries throughout the autumn of 1783, including open clusters in the region of the constellation Cassiopeia. In her diary, she noted that by the end of 1783 she had found fourteen new objects. Just a few months after giving up her career in music, she

Charles Messier

The name Messier is well known to astronomical observers, from novice to professional. The catalog of deep sky objects compiled and published by French astronomer Charles Messier provides the starter-set of objects beyond our solar system to observe with a telescope.

Messier was born in 1730, and he developed an interest in astronomy at a young age. In 1758, while searching for Comet Halley, he noticed a faint smudge of light in the constellation Taurus. Since this object was not noted in any catalog of stars or nebulous objects, he thought it might be the comet. But unlike comets, the object did not shift against the background stars from night to night, and he noted it as the first object in his famous catalog of objects not to be mistaken for comets.

Over the next three decades, Messier scanned the skies for comets and turned up other nebulae in the process, and in 1771 he published a list of forty-five nebulous objects. Two supplements published over the next decade brought this list to 103 objects. During the last century, historians working with Messier's observing notes have expanded this list to 110 entries.

Most of the objects on Messier's list are not nebulae at all, but with the small refractors he used he could not resolve many of his objects to the level of detail to see their true nature. The catalog contains fifty-six star clusters, forty galaxies, six emission/reflection nebulae, four planetary nebulae, one supernova remnant, and three other objects.

had, independently, increased the number of nebulae known to the astronomers of her day by more than 10 percent. Astoundingly, she had accomplished this not with one of the giant reflectors that had helped her brothers build, but with a tiny refracting telescope that some would later refer to as a toy.

FIVE

Star Clusters and Mysterious Clouds

aroline's discoveries of nebulae were a turning point for her as an astronomer and lifted her self-confidence. She decided to put together a star catalog, her first independent research project. His sister's dedication and success impressed William so much he built her a Newtonian reflector of two feet focal length, with an aperture of 4.2 inches, and a magnifying power of twenty-four. It was designed to be small enough for Caroline to use while seated, with her eye at rest in a comfortable position as she swept the sky, from zenith to the horizon.

Caroline's success made William realize that examining nebulae was a fruitful enterprise. He decided to take over this arena himself. He would try to discover the nature of these objects and classify them. If Caroline was upset that her brother snatched away her opportunity to discover new nebulae she never mentioned it.

Their twenty-foot telescope with its twelve-inch mirror was still mounted very crudely and was difficult to use. William designed another, similar telescope—another twenty-foot scope with an eighteen-inch mirror. This time he paid attention to designing the telescope's mounting to make sure it was stable for the systematic scrutiny of nebulae and other stellar objects.

Telescopes with moderate magnification capabilities were suitable for comet-hunting, the goal of which was to review the entire hemisphere rapidly, searching for new appearances before they were spotted by someone else. In contrast to the hurried observations of a comet-hunter, someone mapping the stars and studying the nature of stellar objects and nebulae had more time. Their positions were permanent and there was no urgency to study them quickly because contemporary astronomers were preoccupied with other questions.

William, however, was impatient to begin. He started his observations with the new telescope before the mounting was completed. Caroline watched with consternation:

> My Brother began his series of Sweeps when the Instrument was yet in a very unfinished state, and my feelings were not very comfortable when every moment I was alarmed by a crash or fall, knowing him to be elevated 15 . . . feet on a temporary cross beam instead of a safe Gallery. The ladders had not even their braces at the bottom and one night in a very high wind he had hardly touched the ground before the whole apparatus came down.

When the telescope was completed, William gave Caroline a task that revealed how little he took into account his or her well-being during the pursuit of knowledge. He decided that she should re-examine double stars that he had previously studied, using the smaller twenty-foot telescope. This meant that she would have to climb a ladder in the dark and use the unwieldy instrument to try to locate the double stars to measure their relative positions and also record her observations. It was impossible for anyone so precariously perched to accomplish much of value.

William also had set himself an unreasonable goal. To observe nebulae, which were faint even through his giant instruments, he had to wait until his eyes were dark-adapted, then tilt the telescope to a chosen elevation, move it a few degrees, and then take his eyes off the eyepiece to record his observations. Once he had written down all that he remembered seeing, he had to wait until his eyes were again adapted to the dark. Then he would repeat the procedure, with the telescope pointed at a different part of the sky. He soon realized that he could not progress this way. To make headway, he needed more than his powerful telescope, he needed a capable partner. It struck him that it would be far more efficient if he could team up with Caroline—she could record his observations, while he peered through the eyepiece, keeping his eyes on the telescope so that they were dark-adapted the entire time. Her records would also be far more dependable and accurate than any William might make on his own, which would have to rely on his memory.

This working arrangement was the core of the twenty-year partnership between the Herschel siblings that began on Dec 13, 1783. During the three winters that William and Caroline spent in damp and dismal Datchet, she became indispensable. As he peered through the telescope, she sat at a table at an open window, surrounded with several lights, with the catalogs gathered by John Flamsteed, the first Astronomer Royal, whose work was critical to Newton's discoveries, open before her. An ink pen was in her hand, though some nights were so cold the ink froze in the inkwell. She watched the hands of the pendulum-clock to note down the time, waiting to hear her brother's instructions, while watching an especially created dial she kept beside her. The hand of this dial communicated by strings with William's telescopes, and provided more information, by signs they had agreed upon, about his observations.

Sidereal time

Caroline used two clocks for solitary sweeping: the sidereal clock and the "Monkey clock." Time is normally measured based on the rotation of Earth with respect to the sun. Sidereal time is measured by the rotation of Earth relative to distant stars instead of the sun. One sidereal day, approximately twenty-three hours and fifty-six minutes, is the time taken for Earth to rotate once with respect to the stars, just as a normal day of twenty-four hours is roughly the time it takes Earth to complete one rotation around the sun. These two times are not equal because the earth is moving around the sun as it rotates on its axis, so that in the time it takes to complete one full rotation (the sidereal day), we have moved in our orbit to a different vantage point with respect to the sun.

The four extra minutes are the time it takes the earth to spin to compensate for the change in perspective and bring the sun through its apparent circuit in the sky. Sidereal clocks, which astronomers use to aid them when they make observations, measure sidereal time. The monkey clock was a loudly ticking metronomic device Alexander Herschel made for Caroline. She used to take it on the roof when sweeping for comets to count seconds by.

When William spotted a nebula, he shouted the description and details. On a large chart, she marked the objects which he enumerated or discovered within particular constellations. She wrote down the celestial latitude and longitude of an object's position (called the declination and right ascension respectively), the angle of elevation of the instrument, and everything else that was required. She then repeated what he had said to him, for verification. With Caroline taking notes for him, William was able to work much faster than he could on his own.

Caroline's own "sweeping was interrupted by being employed to write down my brother's observations with the large twenty-foot" and she "became intirely [entirely] attached to the writing-desk, and had seldom an opportunity after that time of using my newly acquired Instrument." This statement was the closest she came to a murmur of complaint at having been forced to give up searching for comets. With characteristic self-abnegation she added that she "had, however, the comfort to see that my brother was satisfied with my endevours [endeavors] in assisting him when he wanted another person either to run to the clocks, writing down a memorandum, fetching and carr[y]ing instruments, or measuring the ground with poles . . . of which something of the kind every moment would occur."

William began conducting sweeps by pointing the telescope at one part of the sky and allowing Earth's motion to bring different parts of the sky into view. This way, each zone (strip of sky) with the same angular distance from the North Celestial Pole could be observed as the movement of the earth—the apparent drift of the heavens above—brought different objects into the telescope's focus. The North Celestial Pole is the point in the sky directly over the earth's North Pole. The stars appear to rotate about this point, because it is the extension—or projection—of Earth's rotation axis onto the Celestial Sphere. The rotation was slow enough that the telescope could be moved up and down slowly, so that each circular strip of sky examined was at least a few degrees wide.

On New Year's Eve of 1783 Caroline was preparing as usual to execute William's instructions. The night started out cloudy but at about ten o'clock the skies slowly began to clear. Impatient to get started, William told her to make some adjustments to the telescope. She set out to do his bidding, but soon an accident occurred:

> At each end of the machine or trough was an iron hook, such as butchers use for hanging their joints upon, and having to run in the dark on ground covered a foot deep with melting snow, I fell on one of these hooks, which entered my right leg about 6 inches above the knee, my brother's call, 'Make haste!' I could only answer by a pittiful crey [pitiful cry], 'I am hooked!' He and the workmen were instantly with me, but they could not lift me without leaving nearly 2 oz. of my flesh behind. The workman's Wife was called but was affraid [afraid] to do anything, and I was obliged to be my own surgeon.

Caroline merely tied a kerchief around her wound for a few days "till Dr. Lind hearing of my accident brought me ointment and lint and told me how to use it." He also informed her that if a soldier had met with such an accident he would have been entitled to six weeks of rest in a hospital. Caroline limped doggedly to her post by her brother's telescope every night clear enough for observation, noting on the night of the incident that she was comforted by the thought that "my Brother was no loser through this accident . . . for the remainder of the night was cloudy."

The incident proved that Caroline's worries about potential accidents were well-founded, but it did not lessen her new found dedication to astronomy. She passed over similar incidents with a casual reference: "I could give a pretty long list of accidents which were near proving fatal to my brother as well as myself," but rather than list them, she pointed out that when she was at her post, "personal safety is the last thing with which the mind is occupied." One of the only other accidents she described in some detail occurred years later, in 1806, when a beam broke and both her brothers "had a narrow escape of being crushed to death."

William stopped sweeping each day with the first rays of daylight but Caroline still had work to do. Each morning she had to produce a copy of the observations they had made the previous night. As the observations were dictated by William and scribbled down by Caroline, they were not organized in any particular way at first. At the end of each night, there would be a long list of observations that Caroline had to organize in a way that was meaningful, arranging the data in tables or cataloging concisely. By the time she was finished, the raw data was organized and distilled into a few key numbers that were easy to use and examine. Caroline did all the laborious calculations and data reduction, all the record keeping, and every other tedious task that required a trained mind. From Caroline's summaries, William could elucidate patterns, make analytical deductions, and formulate theories.

Caroline also planned the schedule of their nightly observations during the day. This way she was prepared beforehand with information about the stars they would see that night. This was not an easy task. She was still relying on Flamsteed's catalog in which the stars were arranged by constellation—not in the order William would see them as he examined zones in the sky. Caroline decided to compile her own list of stars. She calculated their positions and

John Flamsteed, the first Astronomer Royal for the Royal Society

ordered them according to the sequence in which they would encounter them. She reorganized Flamsteed's British catalog into zones of one-degree width so that she could prepare, in advance, a list of what William could expect to find, as they searched the skies systematically.

Caroline also began preparing a catalog of all the objects she and William were discovering. This work would one day become the basis of the New General Catalog, an astronomical catalog still in use today. When Caroline and her brother began their survey, there were about a hundred recorded nebulae in Messier's catalog; by the time they finished, they had expanded the list of known deep sky objects to 2,500. Astronomers could no longer afford to be myopic and study just the solar system—the sheer magnitude of the Herschels' discoveries broadened the field of astronomy.

Late in the spring of 1784, as William was shouting out counts of stars that drifted into his field of view, he suddenly became quiet. Finally, after several moments of silence, Caroline heard her brother exclaim with excitement, "Hier ist wahrhafting ein Loch im Himmel [Here indeed is a hole in the heavens]!" William had come across a dark nebula— matter that obscured his view. He made a passing mention of this discovery in a subsection of a paper he presented to the Royal Society a year later, conjecturing that there was some sort of absence of objects in that part of the sky. Caroline, however, had a hunch that this finding was more noteworthy, that it was not just a hole in the heavens. Fifty years later, she brought these mysterious objects to the attention of John Herschel, William's son. Around that time, she also compiled the first-ever catalog of dark nebulae, which she extracted from the large body of observations that she and William had made. The significance of this portion of Caroline's work was made clear only in the early twentieth century, when astronomers demonstrated conclusively that these were not vacancies but actual astronomical objects that contained dust that obscured the view.

Dark Nebulae

William Herschel's "hole in the heavens" was the first documented telescopic observation of a dark nebula. But with his description of the region as a hole—as devoid of stars—he established an idea that was difficult to shake from the minds of astronomers.

For more than a century astronomers clung to the idea that the dark and relatively starless patches of sky were windows into deep space that let us peer beyond the confines of our Milky Way.

Just over a century after William called his "Loch im Himmel" down to Caroline, Edward Emerson Barnard was applying the techniques of photography to his astronomical studies. Barnard was an accomplished photographer and an extraordinary observer with a reputation rivaling Herschel's. He began taking large scale photographs of the skies in 1788, and noticing dark starless regions in his pictures. He continued this work over the next three decades, but throughout this time his thoughts about the matter were influenced by Herschel's description of such regions as holes.

In a flash of insight, Barnard solved the puzzle of the dark nebulae. He recalled once observing from a remote and dark location on a clear moonless night. A few tiny puffs of cumulus clouds drifted overhead, obscuring the stars as they passed. The appearance of the shifting clouds was as holes in the heavens—just like Herschel had seen. The dark nebulae were clouds of interstellar matter between us and the distant stars.

We now know that the dark nebulae are large clouds of interstellar gas and dust. The dust particles block light from passing through.

The dust also shields the interiors of these clouds from radiation, allowing the cloud centers to become very cold. Under these conditions, the clouds can collapse to form stars. Dark nebulae are often found near emission and reflection nebulae which have been illuminated by stars that recently formed out of the deep dark clouds of gas and dust.

Rho Ophiuchi, William's hole in the heavens

When the seasons changed and summer approached, Caroline used the leisure hours of the lengthening day to help design and build telescopes. The Herschels' telescopes had become famous; scientists, amateurs, and royalty vied with one another to purchase them. King George III ordered several, some of which were given away as gifts.

Astronomer Sir William Watson visited the siblings frequently at Datchet. He became dissatisfied with their financial situation. William wanted to make improvements to his instruments and to build better ones, but the salary from the king did not arrive as regularly as had been promised. He had to dip into his own scanty savings in order to maintain his existing telescopes and to construct the larger telescopes that he had planned, and on which his career now depended. At Bath, Watson often met people belonging to the Royal Court, and he did his best to publicize William's monetary difficulties amongst these wealthy folk. As a result of Watson's efforts, a rich gentleman named Sir Joseph Banks promised "that 2000 would

Wealthy explorer Joseph Banks helped fund William's telescope building.

be granted for enabling him to make himself an instrument. Immediately, every preparation for beginning the great work commenced."

Datchet was an unsuitable location for the new giant telescope construction project. The ceiling still leaked and the dampness made William come down with a fever. In June 1785, William and Caroline moved to a home called Clay Hall, in Old Windsor. Unfortunately, their new landlady turned out to be a "litigious woman," and on April 3, 1786, brother and sister moved to Slough, not far from Datchet. By now, Caroline was an expert at relocating their establishment: "among all this hurrying business, every moment after daylight was allotted to observing. The last night at Clay Hall was spent in sweeping till daylight, and by the next evening the telescope stood ready for observation at Slough."

It took some time for observations to start running smoothly at Slough, but soon they were hard at work. Caroline remarked:

Caroline and William briefly lived at this house, called Clay Hall. (*Courtesy of Science & Society Picture Library*)

"If it had not been sometimes for the intervention of a cloudy or moonlight night, I know not when my Brother (or I either) should have got any sleep; for with the morning came also his workpeople, of which there were no less than between thirty or forty at work for upwards of three months."

The washhouse at Slough was converted into a forge for manufacturing tools, and Caroline's attitude towards the "complete workshop for making optical instruments" was one of enthusiasm and excitement. No longer upset by the workshop, as she had been during her days at Bath, she now said: "it was a pleasure to go into it."

The barn at the Herschels' home in Slough, where William constructed his forty-foot telescope. *(Courtesy of Science & Society Picture Library)*

The move and all the construction activity created some confusion that Caroline later remembered "with regrett [regret] that . . . so much labour and expense should have been thrown away on a swarm of pilfering workpeople. . . . I required my utmost exertion to rescue the manuscripts in hand, from destruction by falling into unhall[o]wed hands, or being devoured by mice."

Caroline took charge of the situation so that her brother was not distracted. By shouldering such a large burden of responsibility, she enabled him to devote all his energy to astronomy.

By 1786, Caroline had made significant progress on her catalog. She recorded: "All the Neb [Nebulae] are registered in Fl[amsteed's] time and P[olar] D[istance] as far as the single nebula in 572 Sweep. The number of the Neb is 1567."

In July that year, one of the ten-foot telescopes King George had ordered was completed. The king wished to present it to the University of Goettingen in Germany, which had been founded by his grandfather (who like him had been both king of England and Elector of Hanover) and ordered William to deliver it in person. William and Alexander decided to also visit their relatives still in Hanover. Caroline was left at Slough with Alexander's wife to take care of the telescopes and to attend to royal visitors and guests.

Even in William's absence, distinguished visitors continued to make their way to Slough. Caroline spent the days calculating and completing her catalog, unless she was interrupted by visitors or by Alexander's wife, who Caroline viewed more as a hindrance than an asset because she "was obliged frequently to sacrifice an hour to her gossipings."

Caroline devoted every clear night to sweeping, enjoying the opportunity to conduct her own work uninterrupted.

On the evening of August 1, 1786, Caroline waited as usual for the sky to darken. About half-past nine, she climbed to the flat roof of the converted stables, known as Observatory Cottage, to begin sweeping in the neighborhood of the sun. She was looking for comets. Around ten o'clock, she noticed something interesting, an object that in color and brightness resembled the 27th nebula in Messier's catalog, but slightly more rotund. It looked like a star out of focus, while the others were perfectly clear. Its fuzzy appearance made her suspect that it might be a comet, but before long, a haze descended on the Thames valley, preventing further observation. She recorded the object's position, drawing familiar stars within the same field of view. It lay in a triangle between stars in the constellations of the Great Bear and Berenice's Hair. With quiet understatement, she wrote in her journal: "I have calculated 100 nebulae today, and this evening I saw an object which I believe will prove tomorrow night to be a Comet."

The skies poured with rain the next day, and Caroline feared that it would be far too cloudy for her to observe that night. She waited nervously until the skies cleared an hour after midnight. She was able to verify that the indistinct object had moved relative to stars. Comets change position from night to night, and because they move so fast in their orbits around the sun their motions can be detected by nightly observations. Planets also appear to move against the background of the stars, but usually more slowly than comets, while stars, in comparison, seem to be fixed in place. Proudly but succinctly, she recorded her comet at one o'clock in the morning with the words: "the object of last night *is a Comet.*"

She did not go to bed that night until she had written to Dr. Blagden, the secretary of the Royal Society, and enclosed drawings indicating the object's position. Her letter, dated August 2, 1786, began cautiously:

SIR, –
In consequence of the friendship which I know to exist between you and my brother, I venture to trouble you, in his absence, with the following imperfect account of a comet.
The employment of writing down the observations when my brother uses the twenty-foot reflector does not often allow me time to look at the heavens, but . . . I found an object . . . suspected it to be a comet . . . it was not possible to satisfy myself as to its motion till this evening. . . . These observations were made with a Newtonian sweeper of 27-inch focal length, and a power of about 20. The field of view is 2°12 . . . whence the situation of the comet, as it was last night at 10 h 33, may be pretty nearly ascertained.

She also wrote to Alexander Aubert, another friend, asking him, self-deprecatingly, to "excuse the trouble I give you with my wag [vague] description, which is owing to my being a bad (or what is better) no observer at all. For these last three years I have not had an opportunity to look as many hours in the telescope." Along with the letter she sent an accurate and detailed explanation of the object's location, and ended her letter with the words: "Lastly, I beg of you, sir, if this comet should not have been seen before, to take it under your protection."

After a few hours of sleep, she went to see Dr. Lind, her physician, who, along with a Neopolitan-born physicist and amateur musician, came to Slough to confirm her observation. Unfortunately, it remained cloudy all that night and the next.

Caroline had done all she could. She waited to hear if she had been the first to spot the comet, or if someone had seen it before she had.

SIX

Days Rich in Discovery

A letter, dated August 5, 1786, arrived from Dr. Blagden informing Caroline, "I believe the comet has not yet been seen by anyone in England but yourself." On the sixth of August he came to offer congratulations and to see the object through Caroline's telescope. Two days later, an elated Alexander Aubert wrote. After apologizing for the delay in his reply, which had been caused by his desire to "give you some account of your comet before I answered it," he continued, "I am more than pleased than you can well conceive that *you* have made it, and I think I see your *wonderfully clever* and *wonderfully amiable* brother, upon news of it, shed a tear of joy. You have immortalized your name, and you deserve such a reward from the Being who has ordered all these things to move as we find them, for your assiduity in the business of astronomy, and for your love for so celebrated and deserving a brother. . . . I found it immediately by your directions; it is very curious, and in every respect as you describe it . . . it travels very fast."

Artist's rendition of a later comet

The news of Caroline's discovery spread quickly. Soon she was famous among Europe's learned people. The "Lady's comet" caused something of a sensation that spread from England through to continental Europe and attracted a steady stream of visitors to Slough.

Caroline's later actions showed that she was proud of her discovery, but at the time she remained modest. The publicity she received did not distract her from her work. She continued to work on her catalog. On August 9 she recorded that she had calculated one hundred nebulae, on the tenth another hundred. On the eleventh she reached a milestone by completing her catalog of the first thousand nebulae, but she did not rest. The next day she calculated "200 nebulae

A drawing of William Herschel's forty-foot telescope(*Library of Congress*)

of the second thousand," and two days later, two days before William returned from Germany, she "calculated 140 nebulae to-day, which brought me up to the last discovered nebulae, and, therefore, the work is finished."

That summer the usually manicured lawn was a chaotic mess as work on the "Giant Telescope" progressed. William had designed the telescope in detail. The supporting stand was made of timber, using a method that later became known as diagonal bracing. The tube was one-twentieth of an inch thick, cast from iron, and made rustproof by a process that was later used for corrugated iron roofing. It was wide enough to allow women to parade through it in their largest feathered hats and widest hoops. Caroline noted with amusement that, "before the optical parts were finished, many visitors had the curiosity to walk through it, among the rest King George III, and the Archbishop of Canterbury, following the King, and finding it difficult to proceed, the King turned to give him the hand, saying, 'Come, my Lord Bishop, I will show you the way to heaven!'"

Time went by "in a perfect chaos of business. The garden and workrooms were swarming with laborers and workmen, smiths and carpenters going to and fro between the forge and forty-foot machinery . . . whilst iron work for the various motions was being fixed." The Herschels were, as usual, impatient to begin observing with the new telescope and started work before everything was in place. Caroline sat at a shelter at its base, recording her brother's observations, while he sat one hundred and fifteen feet above, dictating through a pipe. When there was too much moonlight for careful observation, they spent the nights conducting experiments.

Caroline made copies of no less than seven of William's papers in time to be delivered to the Royal Society between 1786 and 1787, although she "very seldom could get a paper out of his hands time enough for finishing the copy against the appointed day for its being taken to town." Caroline also:

> had always some kind of work in hand with which I could proceed without troubling him with questions; such as the temporary index which I began in June 1787. Some years after, the index to Flamsteed's observations, calculating the beginning and ending of the sweeps and their breadth, for filling up the vacant places in registers, and works

of that kind, filled up the intervals when nothing more necessary was at hand.

When the telescope was completed, there was a gala inaugural, at which Caroline "was one of the nimblest" and the "foremost to get in and out of the tube" as she led the rejoicing cavalcade in a dance through it. Her nephew George Greisbach, a member of King George's band, played the music at the celebration.

The inauguration of the great telescope attracted even more visitors, but guests had ceased to be Caroline's chief worry. There was cause for greater concern.

The house at Slough belonged to a lady whose son-in-law, John Pitt, and daughter, Mary Pitt, had become William's close friends. In the summer of 1786, John Pitt died, and William continued to visit the widowed Mary. William and Caroline often walked over to Mary's house to console and comfort her. William also often invited Mary over to Slough. Soon, it became clear that William wanted to make Mary his wife.

Caroline was shocked. She knew her brother was considered a highly eligible bachelor, but he had not shown any interest in marriage. At Bath, William had had many female friends, but to her knowledge, he had never been attracted to any of them. Now, he was fifty-years-old and she had assumed that she would never need to relinquish her place of paramount importance in his life.

When William asked for Mary's hand he suggested that the couple live at Mary's home, and that Caroline remain at Slough, which would remain his house of work. Mary did not want to accept these conditions. But William seemed unwilling to compromise and it looked as though his engagement

William Herschel *(Courtesy of Art Resources)*

to Mary would be sacrificed. But William and Mary soon negotiated terms they could both agree to. There would be two households, Mary's and William's. Caroline would remain William's astronomical partner, but she would live in an apartment on the grounds of Slough, over the workshops.

Caroline was crushed. For more than a decade she had been first in her brother's affections and had presided over his household. She had also "been almost the keeper of my brother's purse, with a charge to provide for my personal wants, only annexing in my accounts the memorandum *for Car.* to the sums so laid out." Caroline had always been careful with money, and especially frugal with her own needs, spending hardly seven or eight pounds on herself each year—but she had been able to take whatever she needed, whenever she needed it, much as a wife might have been able to do. Now this would change. She had sacrificed her career, deflected praise, and worked solely and devotedly for and with William, in exchange for little more than a roof over her head. Now, the man who had been the center of her life and around whose needs her entire world revolved, was about to step into a sphere of life in which she had no part.

William offered to provide Caroline with a monthly sum to cover her personal expenses, but Caroline saw this as dangerously similar to her early days of servitude to her eldest brother Jacob. She did not want to have to survive on handouts from a married brother. Though it was accepted practice at the time, in her mind it differed little from charity.

For once in her life, Caroline disagreed with William and made her desires clear. She wanted no more and no less than her financial independence, and was prepared to fight for it. She declined her "dear brother's proposal (at the time he resolved to enter the married state) of making me independent, and desired him to ask the king for a small salary." Caroline demanded payment from the king himself for her astronomical work. Her brother was receiving royal remuneration for his work and she wanted compensation as well, she argued. Reasonable though this request seems, it was unprecedented. No woman had ever received a regular salary for scientific work.

William agreed to Caroline's wishes. In any case, he had to write to the king again requesting more money to maintain the massive forty-foot telescope. The construction and maintenance were considerably over budget. At the same time he asked that Caroline be given a lifetime pension of about fifty pounds per year, a quarter of his own salary. He justified this request by saying that his good sister worked harder and more to his liking that any other person he could find. Perhaps the gracious queen, William suggested, by way of encouraging a female astronomer, could be induced to award Caroline this sum, so as to relieve Caroline's anxiety of being left destitute should anything happen to him. William also assured the king that if Caroline were replaced by someone else they would have to pay twice as much in wages.

The king thought he had already finished paying for the telescope and was irritated at William for asking for more then he had initially stipulated. He flew into a rage, but luckily for the Herschels, after his fit subsided, he conceded to William's request.

Caroline was officially awarded the title of Assistant Astronomer to William. The king agreed to pay her an annual sum throughout her life of fifty pounds, which Caroline observed was "exactly the sum I saved my Brother at Bath in writing music." Caroline felt somewhat degraded, however, because the king made it very clear that this was to be the final request and that William was not allowed to ask for any more money. In her opinion, the king never gave her brother as much as he deserved, and the support he did provide was not granted in a gracious manner.

In the spring of 1788, William married Mary. Caroline and Alexander were the witnesses, and Caroline supervised the wedding breakfast, despite her resentment against her sister-in-law. In the only surviving paragraph hinting at her displeasure, she wrote: "The Cat. [catalog] of the second 1000 new Nebulae wanted but a few numbers in March to being complete. And the observations on the Gregorian satellites furnished a paper which was delivered to the Roy. Soc. [Royal Society] in May. The 8th of that month being fixed for my brother's marriage; it may easily be supposed that I must have been fully employed (besides minding the heavens) to prepare everything as well as I could against the time I was to give up the place of a Housekeeper, which was the 8th of May 1788."

Caroline restricted expressing her despondency to her diary. She later destroyed the ten years of her entries written during the period surrounding William's marriage. No record written in her hand provides proof of her jealousy towards her sister-in-law. However, decades later, she broke off her autobiography abruptly around the time of William's wedding and barely referred to the event at all, which suggests that she never fully recovered.

Caroline's emotional turmoil was alleviated somewhat by the arrival of her first salary payment: "in October I received twelve pounds ten, being the first quarterly payment of my salary, and the first money I ever in all my lifetime thought myself to be at liberty to spend to my own liking. A great uneasiness was by this removed from my mind."

After the wedding, Caroline moved into the apartments at Observatory Cottage, where she soon learned the reality of being dependent on the king's generosity. Despite his commitment, the king stopped her salary. She wrote to William, who had the necessary tact to remind the king to honor his promise, and payments started to appear regularly.

The payment of this salary was more than a personal victory and the long-awaited achievement of a personal dream of financial independence. It was a landmark in the history of science. No European woman had ever before received such a high level of official recognition for her scientific contributions and been given a paid scientific office by a government. Caroline Herschel had become the first professional female astronomer in recorded history.

Caught between day and night, Comet Kohoutek streaks through the sky in this image.

SEVEN

Most Admirable Lady Astronomer

Caroline's routine changed little after William's marriage. Although she saw "his domestic happiness pass into other keeping," she continued to work as hard as before "with saddened heart but unflagging determination."

William also remained dedicated to his work, but marriage did cause him to slow down somewhat. He and Mary often vacationed together. Caroline never joined them—it did not occur to William to ask her to come along. When he went on holiday, he left her behind to continue observations and to entertain the steady stream of visitors. As he toured the countryside, he would send her long lists of instructions:

> I shall be glad to know the names of every person who has called at Slough either to inquire after me, or to see telescopes; and also to hear whether more papers have been sent to Bulmer for correction. If he wants to know what copies I wish to have, you will say as usual that Dr. H. wishes to have 25 copies. Perhaps Dr. Davy may write

something about the engravings of Nebulae; of course you
will mention that I shall be at home in about 6 weeks and
will inspect the correction of the plates myself.

Caroline was not always alone when William was absent.
Alexander's wife died the same year William married and her
widowed brother was often at Slough to keep her company.

As the year 1788 drew to a close, Caroline astounded the
male-dominated astronomical world once again by discovering
another comet, proving that her first discovery had not been
a lucky fluke. She spotted her second comet using the small
Newtonian sweeper. This time, William was nearby and could
confirm her sighting with his ten-foot Newtonian. In a short
letter to Maskelyne, she wrote:

> Dear Sir-
> Last night, December 21st, at 7h 45, I discovered a comet,
> a little more than one degree south—preceding Lyrae.
> This morning, between five and six, I saw it again, when
> it appeared to have moved about a quarter of a degree
> towards of the same constellation. I beg the favour of you
> to take it under your protection.
> Mrs. Herschel and my brothers join with me in compliments
> to Mrs. Maskelyne and yourself, and I have the honour
> to remain,
> Dear Sir,
> Your most obliged, humble servant,
> CAROLINA HERSCHEL.

William added a postscript, describing the comet's position
in slightly greater detail. He also wrote the same day to Sir

The seventh telescope William made for Caroline *(Courtesy of Science & Society Picture Library)*

H. Englefield, an astronomer who had drawn up tables predicting the position and time of return of a comet that had been sighted in 1661. In his letter, William mentioned that he and his sister had been unable to find the 1661 comet—but Caroline had found a new one.

Sir Englefield's gracious reply was dated December 25, 1788, two days before Maskelyne's. He wrote to William, "I am much obliged to you for your account of the comet, and beg you to make my compliments to Miss. Herschel on her discovery. She will soon be the great comet finder, and bear away the prize from Messier and Mechain."

Maskelyne wrote to Caroline on the December 27, 1788, informing her that he had only received her letter on the twenty-fourth, and that he had delayed acknowledging the letter until he had seen the comet for himself. He went on to explain that although he was almost certain that she had been the first to sight this comet, there was a remote possibility that she might have observed the 1661 comet or one that had been discovered by Messier the previous month.

He ended his letter on a humorous note, saying he

> would not affirm that there may not exist some astronomers so enthusiastic that they would not dislike to be whisked away from this low terrestrial spot into the higher regions of the heavens by the tail of a comet, and exchange our narrow uniform orbit for one vastly more extended and varied. But I hope you, dear Miss Caroline, for the benefit of terrestrial astronomy, will not think of taking such a flight, at least till your friends are ready to accompany you.

Maskelyne credited Caroline as an astronomer in her own right, independent of her brother.

Maskelyne was not the only astronomer to recognize Caroline as her brother's colleague, rather than his assistant. Throughout Europe, male astronomers began to acknowledge that a woman was working in their midst.

Alexander Aubert *(Courtesy of Science & Society Picture Library)*

The Herschels and their achievements were mentioned by their peers in one breath as recognition rose for Caroline's independent achievements as well as her pivotal role in William's work.

Caroline's next comet discovery came on January 7, 1790. This comet was not easy to see, but Charles Messier also saw it and described it as nebulous with bright condensation. It disappeared from view on January 21.

Barely three months later another comet appeared in the sky and Caroline lost no time in finding it. On April 17, 1790, she discovered her fourth comet. When she spotted it, it was not very bright, and had no tail, and she was far

Caroline's second comet was located in the same region of the sky as the Andromeda Galaxy. *(Courtesy of Science & Society Picture Library)*

from elated when she first noticed the elusive object. William was vacationing in Yorkshire and could not help confirm her sighting. She wrote to Alexander Aubert on April 18 confessing that she "did not know what to do" with the comet she had found the night before, because her "new sweeper" was "not half finished."

By the next night, Caroline was more certain of her discovery. In her letter on April 19, she wrote confidently to Sir Joseph Banks:

[it] is a little more than 3 ½° following alpha-Andromedae,

and about 1 ½ ° above the parallel of that star. I first saw it on April 17ᵗʰ, 16 h 24 sidereal time, and the first view I could have of it last night was 16h 5. As far as I am able to judge, it has decreased in P. D. nearly 1°, and increased in A.R. something above 1'.

These are only estimations from the field of view, and I only mention it to show that its motion is not so very rapid.

The reply from Banks was dated April 20, and Aubert's the day after. Joseph Banks commended her for her latest discovery, and assured her that he would "take care to make our astronomical friends acquainted with the obligations they are under to your diligence. I am always happy to hear from you, but never more so than when you give me an opportunity of expressing my obligations to you for advancing the science you cultivate with so much success."

Aubert was unable to see the comet, noting that he had only observed something faint in the region Caroline had indicated. Maskelyne could not find the comet either, although by then Caroline's reputation was so well established that her discovery went unquestioned. Aubert ended his letter by congratulating Caroline, despite his inability to corroborate her observations: "You cannot, my dear Miss Herschel, judge what pleasure I feel when your reputation and fame increase; everyone must admire your and your brother's knowledge, industry, and behavior."

Maskelyne sent a disappointing note to Caroline on April 22, 1790, to let her know that he had still not observed her comet. He pointed out, though, that Caroline's "second communication, at the same time that it gives me fresh spirits as to the certainty of its being a comet, will certainly assist me in more readily finding it. I feared that your using your new telescope might make that a bright comet to you which might prove but a very faint one. . . . As soon as I shall have seen it I will send you a line."

Caroline's sighting was eventually confirmed. By May the comet had brightened, and its tail was stretched across the sky. On June 29, it was sighted for one final time.

Congratulatory letters poured in. Caroline's colleagues were awed by her achievements. She was piling up discovery after discovery, proving to the world they were not chance discoveries, but were the work of a highly skilled professional. Her male counterparts heaped praise upon her. William Wollaston, who had made several advances in designing lenses, acknowledged her as their equal, apologizing for having first put her down as merely a sister astronomer. Maskelyne called Caroline his "worthy sister in astronomy." On July 12, 1790, French astronomer Jérôme Lalande wrote to inform her she now had a goddaughter named after her, because it was impossible "for me to give her a more illustrious name than yours [pour …je ne pouvouis lui donner un nom plus illustre que le vôtre]." To Lalande, she was the "Savante [learned]" woman and he addressed his letter to "Mlle Caroline Herschel, Astronomer Célébre, [celebrated astronomer], Slough." Three years later, Professor Seyffer of the University of Göttingen sent Caroline a paper on the eclipses of the sun. In the accompanying letter he spoke of her in admiring tones— referring to her as "most revered lady" and "noble and worthy priestess of the new heavens." Recalling an earlier visit, he wrote: "I still recall the happy hours passed in England in earlier days of sweet remembrance, and above all, those which I was privileged to spend near you in a society as genial as it was intellectual." He also said that he held her "in the highest esteem."

This esteem increased when, on December 15, 1791, Caroline found another comet, using the large sweeper William had designed for her. Lalande praised and nominated her for a prize that had been established by the French National Assembly for the most useful work, or most important discovery, in the sciences or the arts. Unfortunately, the committee decided to give William

William Wollaston, a renowned lens maker, was among the many distinguished scientists to congratulate Caroline on her astronomical accomplishments. *(Courtesy of Science & Society Picture Library)*

the first prize and decided against giving a second prize to another member of the same family.

William's deserving of first prize was indisputable. While Caroline's peers accepted her as an astronomer in her own right, there were important professional differences between the two Herschels. Both were gifted astronomers, but William was also a theoretician. He was curious about the nature of the universe and innovative—a creative thinker who was able to make the type of scientific leaps that distinguished him as one of the most influential astronomers of his time. He wanted to learn about the nature of each discovery, to understand the physical causes of what he and Caroline observed. He preferred to study the fixed nebulae—his sister preferred catching sight of comets as they blazed through the sky. Her enjoyment came from observation and discovery; his from interpretation. Her love of the science first originated in her devotion to William; his was born in the subject itself.

Although Caroline had achieved an uncommon level of skill at observational astronomy and the transcription and reduction of astronomical data, she stopped short at trying to explain the phenomena she describes and made little effort to learn more about them. Once she discovered her comets, Caroline had no further interest in them. She showed little intellectual curiosity. She was like an explorer whose prime motivation was the discovery of new species or continent. She cared little for formulating theories or speculating on the nature of the universe.

Yet Caroline's achievements in astronomy went well beyond mere observation and discovery. She also undertook, on her own motivation, tasks at which she was gifted. Her work cataloging facilitated the work of generations of future astronomers. Her laborious calculations were also used in the future study of astronomy. While she was not a creative thinker, she still made significant contributions to astronomy.

Caroline added to the great body of astronomical data, but she continued to use William's instructions. Rather than think for herself, she ran to him, reluctant to distance herself from William's supervision.

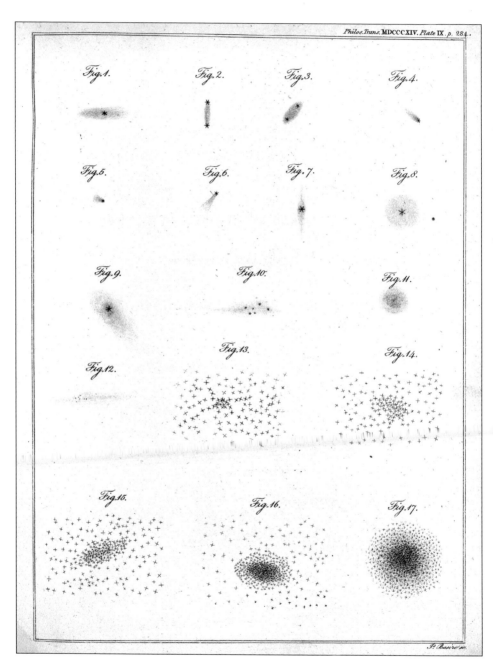

This engraving accompanied an article written by William Herschel in the Royal Society journal in 1814. *(Courtesy of Science & Society Picture Library)*

Caroline also refused to learn more mathematics than she needed to carry out the tasks she was interested in. She was not interested in mathematics for its own sake, but only for applying it to practical problems. To her, mathematics was a tool, not an end in itself, but she had a remarkable ability to use this tool.

Though Caroline "could never remember the multiplication table but was obliged to carry always a copy of it about" with her, the accuracy of her calculations was unparalleled. In 1955, Sister Mary Thomas a Kempis wrote that she could not find a single computational error in Caroline's work, despite the tremendous volume.

The strength of Caroline's calculations lent weight to William's observations and inventive hypotheses. His creativity complemented her aptitude for mathematics. Their teamwork made many of William's advances possible, and the quality of William's work owed much to Caroline's calculations and her attention to detail.

As William began to use the data they had collected to find answers to some of his questions about the universe, Caroline frequently assisted him in writing his scientific papers. Every paper that William published was aided by Caroline in one fashion or another. She did the necessary, laborious calculations, or gathered the paper together out of his disorganized notes, reduced the vast amounts of data they collected, and wrote out a fair copy of each paper before it was sent out for publication. William's letters reveal how hard this job was. He once wrote to her, "My last paper consisted of eighty pages, so that you will have a piece of work to gather it together out of the scraps I leave . . . the rest remains in bits."

EIGHT
Catalog

On March 9, 1792, Caroline's attentions were diverted from the skies when her nephew John, son of Mary and William Herschel, was born. That same year, Jacob Herschel was murdered—but this incident did not seem to have much of an impact on Caroline.

Caroline was an adoring aunt. She nurtured John's interest in science. She was his first astronomy teacher, starting his lessons at an early age, much to his parents' amusement. She encouraged him as he ran about between the great telescopes "in petticoats . . . working with little tools" of his own. She fostered his childish curiosity, and "many a half or whole holiday he was allowed to spend with me, was dedicated to making experiments in chemistry" using boxes, tea canisters, teacups, pepper shakers and other equipment she provided.

The birth of Sir John, as Caroline began to call him, spurred Caroline to try to construct a relationship with her sister-in-law. Mary was of a forgiving nature and happy to build a friendship. Although they never grew close, in the

Caroline's nephew, John, was born in 1792. (*Courtesy of Science & Society Picture Library*)

years after John's birth they developed a lasting affection for one another. Their relationship progressed to such a degree that by 1803, Caroline, who had before detested financial handouts, even agreed to accept a small quarterly allowance of ten pounds from her sister-in-law. This amount was one that Mary and William insisted on adding to her income, as a result of their growing prosperity. Mary was independently wealthy, and her estate had grown since the marriage, thanks to legacies left by her aunt and her mother, until William's royal pension paled in significance.

Though he no longer required extra income, William continued to make and sell telescopes. It was a matter of pride that he could create fine instruments. Caroline assisted in their manufacture and included meticulous notes on their use when they were shipped. The siblings sent telescopes to the Spanish king and the British prime minister, complete with assembly instructions in Caroline's neat handwriting.

When she was not at her desk, the workshop, or by William's side, Caroline used her free time to search the skies for comets, spending many "happy hours . . . on my little roof at Slough." On October 7, 1793, she discovered another one but unfortunately Messier had made the first sighting. However, the letter Caroline wrote to Mr. Planta informing him of her independent discovery of her sixth comet was printed in the *Philosophical Transactions.*

Caroline's seventh comet discovery occurred about two years later, on November 7, 1795. Her letter to Joseph Banks announcing it was calm, revealing none of the agitated excitement of her earlier discoveries. She wrote confidently: "Last night, in sweeping over a part of the heavens with my five-foot reflector, I met with a telescopic comet . . . It will probably pass between the head of the Swan and the constellation of the Lyre, in its descent towards the Sun. The direction of its motion is retrograde."

Caroline did not know when she described the comet's motion that it would become famous. In 1818, the astronomer Encke observed it again, computed its orbit and realized that

PHILOSOPHICAL TRANSACTIONS:

GIVING SOME

ACCOMPT

OF THE PRESENT

Undertakings, Studies, and Labours

OF THE

INGENIOUS

IN MANY

CONSIDERABLE PARTS

OF THE

WORLD·

Vol I.

For *Anno* 1665, and 1666.

In the *SAVOY*,

Printed by *T. N.* for *John Martyn* at the Bell, a little without *Temple-Bar*, and *James Allestry* in *Duck-Lane*,
Printers to the *Royal Society*.

The official publication of the Royal Society

Johann Franz Encke

it was the same comet that Mechain had reported in 1786, and that Caroline had seen in 1795. The discovery of the comet's periodicity was so important that the comet was named Encke's, although years later, John Herschel would call it hers.

Based on William's recommendation, Caroline also took it upon herself to correct the errors in Flamsteed's catalog, which had been considered the astronomical bible until the

Herschels began their observations and noticed it had many errors and omissions. One of the greatest flaws was that the coordinates given in the catalog were often incorrect.

Rectifying Flamsteed's catalog to make it reliable was a huge task. Caroline divided the sky into constellations,

Flamsteed and his Catalog

After he discovered the planet now known as Uranus and attempted to name it after King George III, William Herschel became the king's Royal Astronomer. He was awarded a lifelong pension and occasionally had to provide the royal court with telescopic observations of objects in the heavens.

Meanwhile, there was another important astronomer in England with a similar sounding title: Astronomer Royal. Traditionally, the Astronomer Royal directed the Royal Observatory at Greenwich and oversaw astronomical timekeeping and other observations and calculations of use for trade and military applications.

There have only been fifteen Astronomers Royal in Britain over the past three and a half centuries. During the time that William and Caroline were making their most important observations, Nevil Maskelyne held the post.

The first Astronomer Royal, John Flamsteed, was appointed in 1675. In addition to running the observatory, his primary project during his tenure was the great catalog of stars known as *Historia Coelestis Britannica*. This catalog was quite controversial in its day because Flamsteed feuded with Isaac Newton and Edmond Halley over the details of its publication. Newton pressured Flamsteed to provide him with the observations so that he could test his theory of the moon, but Flamsteed was a perfectionist and resisted letting go of his observations until he was satisfied with them.

Flamsteed carefully measured the positions of more than three thousand stars visible from the British Isles, and numbered them in his catalog according to increasing right ascension (the west-to-east coordinate of the sky) within each constellation. For instance, 1 Tauri would be the westernmost star in the constellation Taurus, 2 Tauri would be the next star to the east, and so on, until all stars within the boundaries of that constellation were cataloged.

Flamsteed's catalog was one of the greatest star catalogs ever assembled, and the Flamsteed numbers are still used by astronomers today. For instance, the star 51 Pegasi was the first sunlike star to have planets discovered orbiting it, and the star 61 Cygni was the first star to have its distance measured through the parallax method.

Caroline made great use of the British catalog when assisting William with his sweeps of the heavens. She rearranged and checked the catalog so that it better suited William's technique of observing, and eventually published her own version of it. Despite Flamsteed's reputation for precision, Caroline noted several errors and omissions in the original catalog.

In a curious connection to the Herschels, Flamsteed actually observed Uranus in 1690 while working on his catalog. He did not recognize it as a planet, as William later did, but logged it in the British catalog as star number 34 in the constellation Taurus.

and within each constellation she listed the stars present. Stars were identified using the number that had been assigned by Flamsteed. By each entry, she also provided the observations Flamsteed had made. Caroline thus created an index by which astronomers could easily cross-reference a star with its description—something that was missing in the original work.

In the course of her research, Caroline also noticed that Flamsteed had left out of his catalog no less than 561 stars.

Caroline compiled a list of these missing stars and also listed the mistakes Flamsteed had made.

In less than two years, working alone, Caroline completed her revision of Flamsteed's catalog. Not only William, but also the Astonomer Royal Maskelyne, recognized the significance of Caroline's hard work. She had put together a reliable and catalog that legions of future astronomers would benefit from.

Upon Maskelyne's recommendation, the Royal Society published her work in the form of a book, with the title: *catalogue of Stars, taken from Mr. Flamsteed's Observations contained in the second volume of the Historia Coelestis, and not inserted in the British catalog, with an Index to point out every observation in that volume belonging to the stars of the British catalog. To which is added, a collection of errata that should be noticed in the same volume. By Carolina Herschel.*

Allowing herself a rare word of praise, Caroline wrote to Maskelyne of her joy that the Royal Society had paid for the publication of her volume: "your having thought it worthy of the press flattered my vanity not a little. You see, sir, I do own myself to be vain because I would not wish to be singular, and was there ever a woman without vanity?—or a man either? only with this difference, that among gentlemen, the commodity is generally stiled ambition."

Although Maskelyne commended Caroline's work and was responsible for its publication, he was disappointed that her list of the 561 missing stars did not provide their actual coordinates. Instead, Caroline had described the position of each omitted star in the context of one of the stars near it. When he told Caroline of his disappointment she began doing

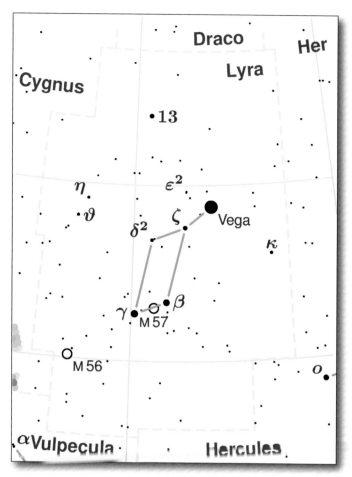

Lyre Constellation Map

the tedious calculations necessary to come up with coordinates for each of the 561 stars, working off a recent catalog prepared by English chemist William Hyde Wollaston. When she sent Maskelyne this revised list he was delighted—and he was not alone. A friend of the Herschels wrote to say that he welcomed her revisions to Flamsteed's opus. "Were Flamsteed alive, how cordially would he thank you for thus rendering the labours of his life so much more useful and acceptable to posterity . . . and future astronomers, as well as those of the present times will doubtless be conscious of the merit and obligation you are entitled to."

Caroline's final comet discovery, like her first, took place when William was away, on August 14, 1797, the same year she completed her revisions to Flamsteed's catalog. In contrast to all the others, which had been spotted through one of her telescopes, she noticed this one with the naked eye. She was looking over the skies early in the evening when she saw it. She listed the comet in her book of sweepings as "C. H.'s" comet.

The delay in the delivery of the letter reporting her second discovery, which took two days to travel from Slough to Greenwich, had given Caroline "little faith in the expedition of messengers of all descriptions." She was afraid someone else would get credit for the discovery first. Instead of waiting, she leaped on horseback after her night's observations were complete and made the twenty-seven-mile ride to the Royal Observatory in Greenwich to report it in person—although she had not ridden over two miles at a time for the past five years—on just one hour of sleep. The effort was worth the pains: her report arrived just hours before another sighting of the same comet by the astronomer Eugene Bouvard.

Within a few weeks of this discovery, for some unknown reason, Caroline decided to move away from her brother's estate and into her own lodgings. She wrote nothing about the reason for her move, and there is no evidence of a rift between her and either William or Mary. Much later, she referred to this period simply as one in which she "came to be detached from the family circle." Around the time that Caroline moved away, Mary and William enlarged the Slough house and made it their sole place of residence, rather than keeping it up in addition to Mary's. Caroline may have moved away because she was discomfited by the idea of continuing to live in the cottage on the grounds of Slough after William and Mary decided to move into the main house.

Whatever the reason, her decision was unfortunate for astronomy. Caroline's new lodgings were far away from the observatory. If she spent a night observing, she had to go home in the dark. The impractical location of her new home was at least partly responsible for a decline in the hours she spent in observation. Caroline herself lamented feeling, "Uncommonly harassed in consequence of the loss of time necessary for going backward and forward, and not having immediate access to each book or paper at the moment when wanted," but she was not one to go back on her decisions. Despite her frustration with her self-imposed decision to move away from the cottage on the grounds of Slough, she did not take up residence there again. Instead, she moved from one unsatisfactory lodging to another for a few years, before finally agreeing to live in rooms at Upton, Mary's former home, where the couple had first set up residence after the marriage.

Caroline never again discovered a new comet—but she had by then established a precedent for female astronomers of the future. She had blazed a new trail by publishing a significant body of scientific work under her own name. In a short space of time, she also made the most comet discoveries of any woman—a record that she held for nearly two hundred years. And although Caroline's days of comet discovery were over, her contributions to astronomy were not. Some of her most important work lay ahead, and one of her most valuable contributions to the field was yet to be made.

NINE

The Passing of William

The Herschels were visited so often by the royals as the years went by that Caroline made only brief records of her contact with the ruling family. "I dined at the castle," she wrote simply, and another time recorded that "the Prince of Orange stepped in to ask some questions about planets." Caroline did not describe her impressions of the famous personages she encountered.

The daughters of King George III, the princesses Sophia and Amelia, were among Caroline's many visitors. Caroline preferred more serious pursuits than entertaining royalty, but she patiently showed the royal princesses views of various celestial objects—Mars, Saturn and its satellites, the mountains on the moon, and the variable star Mira Ceti. Princesses Sophia and Amelia took a great liking to Caroline and questioned her vigorously about various astronomical subjects.

Through her contacts with royalty, Caroline met Madame Beckedorff, one of the ladies-in-waiting to the queen. To her surprise, Madame Beckedorff was an erstwhile

Caroline befriended the Princess Sophia and her sister Amelia.

classmate—she had attended the dressmaking school Caroline had persuaded her mother and brother Jacob to allow her to participate in years before, in Hanover. The women formed a friendship that lasted the remainder of Caroline's life. In Madame Beckedorff, Caroline finally found the type of friend she had long desired: a sincere and sympathetic person, in whom she could confide and find comfort.

Caroline became more sociable than she had been in the past. During the years around the turn of the century she recorded many a pleasant evening at Madame Beckerdoff's home and at the royal residence in Buckingham in her

Caroline was invited to the Royal Observatory by Maskelyne.

diary. In 1799, Caroline was invited by Maskelyne to visit the Royal Observatory at Greenwich.

Though Caroline was an acknowledged astronomer, her brothers still saw her as a housekeeper as well. In 1800, William sent her to Bath for an extended stay to help set up a new house he had bought there. "Began to pack up what I must take to Bath with me, for there I am to go!" Caroline said with exasperation, but to Bath she went.

Caroline's younger brother Dietrich visited England between 1809 and 1813, "ruined in health, spirit, and fortune." During

Caroline's nephew John attended Cambridge University.

his visit, Caroline was forced to work doubly hard because she was expected to continue her demanding astronomical routine, while at the same time, "according tho old Hanoverian custom, I was the only one from whom all domestic comforts were expected." Pointing out that she never neglected the work she did with William, she wrote: "the time I bestowed on Dietrich was taken entirely from my sleep, or what is generally allowed for meals, which were mostly taken running, or sometimes forgotten entirely."

Soon after Dietrich returned to his native Hanover, the Herschel siblings realized that Alexander, who was now a childless widower approaching his seventies, could no longer manage on his own. Caroline offered to care for him, but he declined, deciding instead to spend his last years in

Hanover. Caroline accompanied her brother part of the way on his journey from England to Germany, before he made his last channel crossing.

Between the years of 1809 and 1813, her nephew John entered Cambridge University, and to Caroline's delight was elected a Fellow of the Royal Society within weeks of graduating at the top of his class. He started an illustrious career as Caroline's own career in observational astronomy was drawing to a close. She documented some of her last observations on mountains in the moon in 1811.

William was knighted in 1816, an honor that Caroline and many others felt was long overdue. Shortly thereafter, John became an astronomer, much to the delight of his father and adoring aunt. William's health was failing, but John was in good hands. Caroline, in her seventies, still had sufficient energy to help him learn to sweep the night sky. The records of John's first two sweeps were in Caroline's handwriting. She helped her nephew with undiminished enthusiasm, returning to her position at a desk at an open window after twenty years of absence. But there was a difference—the help she now rendered was in the capacity of a mentor and elder colleague.

In 1818, Queen Charlotte died, and Madame Beckedorff's services were no longer needed at court. She decided to return to Hanover. This was unfortunate for Caroline, who had begun to realize she might outlive William and have to plan for the final years of her life. This was the time that she most needed and wanted Madame Beckedorff's friendship and counsel, but her friend was too far away to help much.

Caroline's existence became increasingly hermetic. She conducted her calculations in her lodgings, except when she

Queen Charlotte of England died in 1818.

"was wanted to assist my brother at night or in his library."
William continued to publish scientific papers with her help,
but his health was worsening steadily. Aware of his declin-
ing health, William urged her to examine "every shelf and
drawer" in his library and workrooms and to transcribe or
make memorandums of all his works. Caroline "found on

looking around that I had no easy task to perform, for there were packs of writings to be examined which had not been looked at for the last forty years. But I did not pass a single day without working in the Library as long as I could read a letter without candlelight, and taking with me papers to copy . . . which employed me for the best part of the night."

William, who was now in his eighties, continued his own observations as best he could with Caroline's assistance. On July 4, 1819, he sent Caroline a note: "LINA,—There is a great comet. I want you to assist me. Come to dine and spend the day here. If you can come soon after one o'clock we shall have time to prepare maps and telescopes. I saw its situation last night—it has a long tail."

Caroline preserved this slip of paper in her diary, with the words: "I keep this as a relic! Every line *now* traced by the hand of my dear brother becomes a treasure to me."

In 1820, Caroline considered returning to Hanover where her brother Dietrich and Madame Beckedorff lived. Certain that Hanover was the only place she could find rest after William's death, she sent thirty pounds to Dietrich to purchase a feather bed for her.

In March 1821, Caroline and William received word that Alexander had died earlier that month. By October, Caroline ended her diary on a note of depression: "Here closed my Day-book for one day passed like another, except that I, from my daily calls, returned to my solitary and cheerless home with increased anxiety for each following day."

Caroline herself fell ill on July 8, 1822, but though she was "seized by a bilious fever," she continued to "rise for a few hours to go to my brother about the time he was used to see me." One day "I was entirely confined to my bed . . . in despair about my own confused affairs, which I never had the time to bring into any order. The next day . . . my nephew . . . promised to fulfil[l] all my wishes which I should have expressed on paper; he begged me not to exert myself for his father's sake, of whom he believed it would be the immediate death if anything should happen to me."

Then, on August 25, 1822, the event Caroline had been dreading came to pass. William, who had rescued Caroline from a lifetime of drudgery, who had expanded her horizons and turned her eyes toward the stars, the man whom she had based her existence around, died in his home at Slough.

Caroline wrote: "not one comfort was left to me but that of retiring to the chamber of death, there to ruminate without interruption on my isolated situation. Of this last solace I was robbed on the 7th of September, when the dear remains were consigned to the grave."

Eighteen months elapsed before, on April 15, 1823, Caroline could finally "acquire fortitude enough to for noting down in my Day book any of those heartrending occurrences I witnessed during the last nine months of the fifty years I have lived in England, and I cannot hope that ever a time will come when I shall be able to dwell on any one of those interesting but melancholy hours I spent with the dearest and best of brothers. But if I was to leave off making memorandums of such events as either affect or are interesting to me, I should feel like what I am . . . a person that has nothing more to do in this world."

Caroline doubted that she had many years of life left: "at that time I should not live a twelvemonth." Stupefied by grief, she made the rash decision to leave England once and for all. Her regular pension had been confirmed by the Crown and she presented every penny of her life's savings to Dietrich. Against John Herschel's advice, she also parted with all that was left of her property. Before she left, John and his wife insisted on giving her part of the yearly allowance which William had provided for her in his will. John had to force her to take the money. After her return to Hanover, John spent many hours pleading with her to use the money that was rightfully hers.

Many of Caroline's distinguished friends, acquaintances, and admirers were dead. Others had moved away. On October 16, a little over a month after William's burial, Caroline's remaining friends assembled at Bedford Palace in London and she took "from all these sorrowing friends and connections . . . an everlasting leave."

TEN

Why did I leave Happy England?

After bidding John farewell in London, but before leaving for Hanover, Caroline used words similar to those she used to describe her childhood loneliness: "I saw no one I knew or who cared for me." This was not entirely true. Though recovering from a serious illness, Dietrich came to fetch his sister. John had wanted to accompany her across, but Caroline's insistence on leaving as soon as possible after William's death prevented it because John had to arrange the funeral and other matters in the wake of his father's death.

Dietrich was no replacement for William. It became clear that in "the last hope of finding in Dietrich a brother to whom I might communicate all my thoughts of past, present and future, I saw myself disappointed. . . . For let me touch on what topic I would, he maintained the contrary, which I soon saw was done merely because he would allow no one to know anything but himself." She realized her "whole life almost has passed away in the delusion that . . . Dietrich was capable of giving me advice." Dietrich complained constantly

This is a photo of London, circa 1900. After William's death, Caroline left her beloved John in London and returned to her birth town of Hanover, Germany. *(Library of Congress)*

about his scanty education—something which Caroline had little sympathy for, having overcome a far poorer education herself; Caroline also felt that this topic insulted the memory of her father.

Hanover was, to her, a disappointment—an "abominable city." In letters to Mary, she complained "What little I have seen of Hanover . . . I do not like! And though some streets

This painting illustrates Hanover in 1895.

have been enlarged (as I am told), they appear to me much less than I left them fifty years ago." Hanover was, "now quite a new world, peopled with new beings, to what I left it in 1772" and she had not considered how much she would have to adjust. She had been away for so long she felt alienated from the town of her birth. "From the first moment I set foot on German ground," she complained to John, "I found I was alone."

Dietrich's wife, whom Caroline first described as an active woman with a cheerful disposition, was pleasant. Caroline also wrote that she was to live in "the handsomest rooms, three or four times larger than what I have been used to . . . prepared for me and furnished in the most elegant style."

A portrait of Caroline's nephew, John, circa 1852. *(Courtesy of Science & Society Picture Library)*

The warm welcome and large rooms did not keep Caroline from missing England. It was "hard to live for months without knowing what may have happened to those with whom one has been for so many years immediately connected and in the habit of keeping up a daily intercourse." She admitted that she could not help "comparing the country in which I have lived so long, with this in which I must end my days, and which is totally changed since I left it, and not one alive that I knew formerly, except my dear Mrs. Beckedorff."

Caroline especially missed helping John with his work. She expressed this regret often: "Believe me, I would not have gone without at least having made the offer of my service for some time longer to you, my dear nephew, had I not felt it would be in vain to struggle any longer against age and infirmity."

Caroline's stubbornness had not softened with the years. Although she was unhappy in Hanover, she refused to return to England. In her mind, she had made a promise never to return, and to leave Hanover now would be breaking her word. In keeping with her character, she also blamed her growing resentment on Dietrich and his family. She did not seem capable of thinking herself even partially responsible for her predicament. She convinced herself that it was they who had deceived her into leaving England by the letters they had written.

Dietrich was a braggart and a spendthrift who quickly fell out of favor with her, but she had made her irrevocable promise, and though she retained her financial independence because of her lifelong royal pension, she had already gifted him with most of the rest of her wealth. When Dietrich fell ill she dutifully nursed him until his death in 1827, but not without frequent complaint.

After he died, Caroline moved into different rooms where she was happier, but her attitude to his widow and his family did not improve. In striking contrast to her initial complimentary account, she wrote an unfavorable description of her Hanoverian reception in her later memoirs. She described Dietrich's wife as a short, corpulent woman, dressed like a girl, her head covered in huge artificial curls, from whose embrace Caroline had shuddered away. She claimed that the rooms which they had set aside for her had had no carpet and no fire and claimed Dietrich's wife was greedy and only came near her when she had some design on her purse. She also said Dietrich's children were no better than his greedy wife. These recollections, made over a decade after her arrival in Hanover, were probably less accurate than her earlier assessment. One of Dietrich's daughters, the widowed Anna Knipping, was a very devoted niece—a fact even Caroline reluctantly acknowledged.

Caroline's increasingly bleak depiction of her life in Hanover was not borne out by John's accounts that were written during his visits to his aunt. She engaged in many activities. Her hectic social schedule included regular evenings with a "learned society, or blue-stocking club," a group of intellectual female acquaintances. Her recently acquired leisure also allowed her to indulge in her long-dormant love of music. Until 1840, she attended concerts and plays regularly, buying herself a season ticket every year for fifteen years.

At the theater, Caroline was a celebrity in her own right who, much to her amusement, was often stared at by the audience. Whenever he saw her, she was "honoured with a wie gehts? [How do you] by His Majesty" the king of Prussia. She was "always sure to be noticed by the Duke of Cambridge as his countrywoman"—much to Caroline's delight because she insisted she was not German. Caroline even discovered that illustrious musicians were as anxious to meet her as she was to hear them. In 1828, she heard the famous Catalani sing, and in 1831, after a concert by

Niccolo Paganini

the violinist Paganini, she found that he had arranged to speak with her through an interpreter.

The pleasure she derived from the theater and other social events did not keep Caroline away from astronomy, which had long since replaced music as her greatest passion. She had brought two telescopes with her—one seven foot, and the smaller of the two sweepers that William had built for her. However, they were of little use to her in Hanover, where the high roofs of nearby houses interfered with her observations of the heavens. Nevertheless, in 1823, she wrote to John that she was "amusing herself with having the seven-foot mounted . . . though I have not even a prospect of a window for a whole constellation." The telescopes stood in her room like monuments, except for rare occasions, when she peeped through them to catch a glimpse of comets discovered by others.

The Hanoverian setting effectively prevented observational work, but there was another astronomical pursuit to which she was still suited, that she turned to as she neared the end of her life.

ELEVEN

Double Stars

In 1823, John Herschel wrote to say that he was about to start a set of observations using his twenty-foot reflector telescope. It dawned on Caroline that despite the distance between them, she could help him in a way no one else could. She could pave the way for John to revise the vast catalog of nebulae and star clusters she and William had accumulated. She could organize the eight volumes containing the data from the sweeps she and William had conducted and revise her initial catalog of star clusters and thousands of nebulae they had discovered. She could also rework the observations she had recorded for William and do all the calculations necessary to find coordinates for the nebulae and clusters and arrange the nebulae she and her brother had discovered, into zones. She would not be at a desk near John, but she had a desk in Hanover that would do.

Having an immense task to work on lifted Caroline's spirits. "I wish to live a little longer," she wrote to John, "that I might make you a more correct catalog of the 2,500 nebulae, which is not even begun, but hope to be able to make it my next winter's amusement." She wanted John to verify and revise, if necessary, the position and description of each of the nebulae and star clusters she and William had discovered.

Caroline had taken the first draft of the catalogs she had made during the twenty-year Herschel partnership. In it the position of each nebula was given in relation to a nearby star. She also had the book of Flamsteed stars she had put together to use in sweeping in which she had calculated star positions for the year 1800, and arranged them in a zone format.

Caroline wanted to produce a catalog of nebulae that matched with the format she had compiled for the stars. It was a complex and complicated undertaking. Because the sweeps overlapped, many nebulae had been seen more than once. She first went through the sweep records and identified each nebula and gave it a serial number according to the chronological order in which it was observed. She noted the date of first observation of each nebula, the star in reference to which William had described the nebula's position, the actual position description in relation to the star, and the reference numbers of the sweeps in which each nebula had been observed. Then she used her star catalog to calculate the coordinates of each nebula. Once the coordinates had been calculated, she grouped the nebulae into zones, based on their angular distance from the Celestial North Pole, as she had done years ago for her star catalogs. Within each zone, she listed the nebulae in the order in which they would appear overhead to an observer examining that zone. The first zone consisted of nebulae lying closest to the North Pole, and within 10° of it; the second zone included all those between

10° and 15°; the third all those between 15° and 17°; the rest of the zones were just one degree wide.

In 1824, John visited his aunt in Hanover and they discussed her progress. In John's opinion she had no cause to regret her change of location—he found Germans in general, and his relatives in particular, to be well mannered, pleasing and peaceful.

Around Easter 1825, Caroline's catalog of nebulae was complete. Sir John Brewster described this as an extraordinary monument of ardor in the cause of abstract science.

As soon as the Zone-catalog was in his hands, John began to review the objects his father and Caroline had discovered decades before. He promised Caroline he would "prize, and more than prize . . . use myself and make useful to others" the work that she had sent him.

John used Caroline's catalog as the basis for part of his own general catalog many years later. He was able to draw up a working list for each night's observations—the type of list Caroline had prepared so she and William would be able to identify every object that floated into their field of view as they swept through each predetermined swath of sky. John wrote that in the course of his work, he "learned fully to appreciate the skill, diligence, and accuracy which this indefatigable lady brought to bear on a task which only the most boundless devotion could have induced her to undertake, and enabled her to accomplish."

John wrote Caroline many letters describing his progress. Caroline, still captivated by astronomy and thrilled to hear the details of how her nephew was following in the footsteps of her brother, regretted she could not "recall the health, eyesight, and *vigor* I was blessed with twenty or thirty years ago; for nothing else is wanting (and that is all) for my coming by the first steamboat to offer you the same assistance (when sweeping) as, by your father's instructions, I had been enabled to afford him."

She could give John other types of assistance, however. She provided hints and instructions on how to proceed and where he might "find undoubtedly many more nebulae which may have been overlooked," helping him to "discover as many objects as would produce a pretty numerous catalog."

In February 1828, Caroline received an unexpected honor for her work when the Royal Astronomical Society voted

Caroline received the highest honor of the Astronomical Society in 1828.

unanimously to award her their Gold Medal. John was the president of the society at the time, but he clarified the decision "was none of my doings." Feeling awkward to speak in praise of his own aunt, he invited James South, the vice president of the society, to announce the council's decision. South's admiration for Caroline was evident in his vice presidential address. He referred to her most recent work, as "the completion of a series of exertions probably unparalleled either in magnitude or importance in the annals of astronomical labour."

Speaking of her contributions as William's colleague, he explained:

> Miss Herschel it was who by night acted as his amanuensis: she it was whose pen conveyed to paper his observations as they issued from his lips; she it was who noted the right ascensions and polar distances of the objects observed; she it was who, having passed the night near the instrument, took the rough manuscripts to her cottage at the dawn of the day and produced a fair copy of the night's work on the following morning; she it was who planned every labour of each succeeding night; she it was who reduced every observation, made every calculation; she it was who arranged everything in systematic order; and she it was who helped him to obtain his imperishable name. . . .
> Indeed, in looking at these joint labours of these extraordinary personages, we scarcely know whether most to admire the intellectual power of the brother, or the unconquerable industry of the sister.

Not surprisingly, at least to anyone who knew her well, Caroline thought South's speech was "clumsy." She was "more shocked than gratified by that singular distinction, for I know too well how dangerous it is for women to draw too much notice on themselves." Caroline's ideas of a woman's duty encouraged her to care more about preserving William's memory than building up her reputation. She explained to John, "Whoever says *too much of me* says *too little of your father*! And can only cause me uneasiness." She constantly deprecated herself, insisting, "he did all," and "I did nothing for my brother but what a well-trained puppy dog would have done: that is to say, I did what he commanded me. I was a mere tool which he had the trouble of sharpening."

Caroline could not conceive of herself as an astronomer in her own right, she saw herself not merely as her brother's assistant, but also as his inferior: "I am nothing, I have done nothing; all I am, all I know, I owe to my brother." As time passed, her adulation for William grew, and her greatest joy came from hearing people praise William, and John's contributions.

Caroline's joy at John's success was tempered by fears he was overworking. She encouraged him to "meet with a good-natured, handsome, and sensible young lady." In 1829, to her pleasure, John married. She offered him a gift of money at his betrothal, which he refused, saying that instead what he really wanted was an oil portrait of her so he could hang it next to one of his father.

A few years after John's wedding, in 1832, Caroline was presented a medal by the king of Denmark, as a tribute for her discoveries. It was John's visit that same year, though, that gave her the most pleasure. John reported that he found his maiden aunt—in contrast to Caroline's self-appraisal— "wonderfully well and very nicely and comfortably lodged, and we have since been on the full trot. She runs about the town with me and skips up her two flights of stairs as light and fresh as at least some folks I could name who are not a fourth part her age. . . . In the morning till eleven or twelve she is dull and weary, but as the day advances, she gains life, and is quite 'fresh and funny' at ten or eleven p.m., and sings old rhymes, nay, even dances! to the great delight of all who see her."

John had exciting news—he was planning an expedition to the southern hemisphere to observe the skies that his father had always wanted to explore. John would now complete the review of the skies William and Caroline had started.

Before John left for the Cape of Good Hope, Caroline wrote excitedly, lapsing into her mother tongue, "Ja! If I was thirty or forty years younger, and could go too? In Gottes namen! [In God's name]!" Caroline, now eighty-two, "jingled glasses with Betty"—her faithful maid—to give vent to her joy. Betty was the sole person who gained Caroline's "entire confidence and approbation" in her final years. Together, they shouted "Es lebe Sir John! [Sir John forever]! Hoch! Hurrah!"

Caroline followed John "in idea every inch" of his journey. She was free with her advice on how to proceed: "As soon as your instrument is erected I wish you would see if there was not something remarkable in the lower part of the Scorpion to be found, for I remember your father . . .could not satisfy himself about the uncommon appearance of that part of the heavens."

John wrote back to say that he had found the region full of beautiful stars, but this was not what Caroline was after. "It is not *clusters of stars* I want you to discover," she replied tartly, "for that does not answer my expectation."

John Herschel sailed to the Cape of Good Hope to observe the Southern Hemisphere.

She demanded that he look for the dark spaces in the sky, recalling William's remark about a hole in the heavens. John wrote again, verifying the existence of dark patches devoid of stars, but ignored her insistence that they were "something more than a total absence of stars." This was the only time that Caroline directly opposed a hypothesis of William's. "I have swept well over Scorpio," she emphasized, bringing to John's attention the "many entries in my sweeping books of the kind you describe viz., the blank space in the heavens without the smallest star." She even copied out locations of dark nebulae—her work was the first to ever catalog such objects—but more than a century passed before astronomers realized that dark nebulae were the birth places of all the stars and planets in a galaxy.

When John arrived at the Cape of Good Hope in 1834 Caroline's large sweeper was with him. He used it to acquaint himself with the unknown southern skies. As he extended William and Caroline's work to the Southern hemisphere,

This photo shows the jet of gas from newly forming Star HH-47. It would be more than a century before scientists learned that William's dark holes were where all the stars and planets of a galaxy were born.

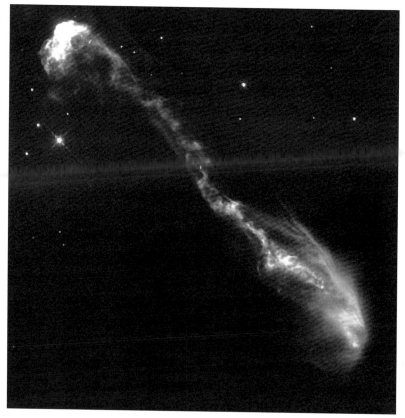

he added close to two thousand findings of his own. All the while, he continued to correspond with his aunt, who described his accounts of work as drops of oil supplying her expiring lamp. Although she "could not help lamenting" that William "could not take to his grave with him the satisfaction I feel at present at seeing his *son* doing him so ample justice," she remarked that John's work had made her "completely happy."

A year later, the Royal Astronomical Society enrolled Caroline, and another woman, Mary Sommerville, as honorary members. Full membership of the society was still barred to women, and in the report to the Fifteenth Annual General Meeting, the society felt obliged to defend its decision to create honorary memberships for them. It pointed out that although "the tests of astronomical merit should in no case be applied to the works of a woman less severely than to those of a man, the sex of the former should no longer be an obstacle to her receiving any acknowledgment which might be held due to the latter."

In 1838, the Royal Irish Academy also named her as an honorary member; the first time such a distinction had been conferred on a woman. Caroline's contributions had won over the prejudices of the time—the person who nominated her, William Rowan Hamilton, was the man who had once refused to allow ladies to attend the academy's meetings.

Caroline's reaction was typical. "I cannot help crying aloud to myself, every now and then, *what is* THAT *for?*" "I think it is almost mocking me," she wrote, "to look upon me as a Member of an Academy; I that have lived these eighteen years (against my will and intention) without finding as much as a single comet." She could not acknowledge that she might

Mary Somerville was made an honorary member of the Royal Society at the same time as Caroline.

deserve the awards for the substantial and significant body of work she had created years before.

At times she poked fun at herself for her whining: "I am the only one who is complaining, but I think I have a right to that preference." When she was felicitated with presentation copies of work done by astronomers of the time, she exclaimed, "Coming to *me* with such things! an old, poor, sick creature in her dotage."

It was not a poor or sick aunt that John found when he visited her that same year, shortly after his return from the Cape.

John brought along his son William James Herschel, which was a surprise for his aging aunt. (He had already named a daughter after her). She was delighted, exclaiming excitedly when she saw him, "Ach mein Gott! Das ist night möglich; ist dieser meines Neffen's Sohn? [My God! This is not possible; is this my nephew's son?]". Then she and her nephew talked "as if we had parted but yesterday." All through the visit she was so concerned about little Willie's welfare that John, hoping to spare them all the sadness of saying good-bye, departed at night, without a proper farewell. Though she realized that the gesture was well meant, Caroline was disappointed. It was the last time they were together.

Caroline retained full possession of her mental faculties as she aged, but as the years progressed she began to find the present somewhat strange and annoying, although she continued to entertain visitors. She was often surprised by letters and gifts from the British royal family, and royalty from continental Europe also kept in touch. Every scientist who visited Hanover paid her a call or at least sent a letter. "Next to listening to the conversation of learned men," she wrote to John's wife, "I like to hear about them."

Her "nimbleness in walking" was admired by all those who knew her, but Caroline pointed out that "the good folks are not aware of the arts I make use of, which consist in never leaving my room in the daytime, except I am able to trip it along as if nothing were the matter."

Sadly, her eyesight began to fail. She wrote that "the few, few stars I get at out of my window only cause me vexation, for to look for the small ones on the globe my eyes will not serve me any longer." Sometimes she roused herself from a "melancholy lethargy" to look over her "store of astronomical and other memorandums of upwards of fifty years' collecting," and destroy "all that might produce nonsense when coming through the hands of a Block-kopff [dunce] into the Zeitungen [newspaper]."

In 1842, Caroline wrote warmly to John's wife Margaret, whom she had never met. "I would not wish on any account to see either my nephew, or you, my dear niece, again *in this world, for I could not bear the pain of parting once more*; but I trust I shall find and know you in the next." She preferred to live with her memories of the past than face the present, but continued to write in her day book until 1845. She began her second autobiography, sent from Hanover to John's wife and children, at the age of ninety-two. She despaired of finishing it before her eyesight and life left her, but could not help writing the history of the Herschels. Her motivation for the work was purely "to be remembered . . . for yet awhile" by John and his family.

On Caroline's ninety-sixth birthday, in March 1846, she received a letter from Alexander von Humboldt informing her that the King of Prussia wished to convey to her, in his name, a "little gratification." It was the large Gold Medal for Science. The king made the award in recognition of the valuable services rendered to astronomy by her as the fellow-worker of her "immortal brother, Sir William Herschel." She was also visited that year by the new Astronomer Royal, George Biddel Airy.

On her ninety-seventh birthday, March 16, 1847, she still had enough energy to entertain her royal guests, the prince and princess, who brought with them a cake and a velvet armchair. She not only carried on a lucid conversation, but also sang them a song that William had composed many years ago when they lived in Bath. Dressed in a new gown and smart cap, she did not appear exhausted when they left, after spending two hours in her company. At the end of the month, she sent a message to her nephew, saying that although she did not quite like old age with its weaknesses and infirmities, she did manage to laugh sometimes, enjoy her meals, and was pleased with the services rendered by her faithful maid Betty. That summer, John sent her his volume—Cape Observations—the completion of the celestial survey the Herschel siblings had begun.

Two months short of her ninety-eighth birthday, Caroline "breathed her last at eleven o'clock . . . the 9th night of January." Inside her coffin, which was adorned with palm branches and wreaths of laurel and cypress sent by the princess from the royal gardens at Herrenhausen, were her father's old almanac and a lock of William's hair. As a mark of respect, the king of Hanover and the prince and princess sent carriages to escort the hearse to the Churchyard of Gartengeinde, at Hanover.

Caroline was buried beside her parents beneath an epitaph she wrote herself. Even in these final words, she reaffirmed her dedication to her brother and her concern that she should shine only in his reflected light:

> The gaze of her who is passed to glory while here below sought the starry skies; of this her own discoveries of comets and her participation in the immortal labors of her brother, William Herschel, bear witness to future ages. The Royal Academy of Dublin and the Royal Astronomical Society of London numbered her among their members. At the age of 97 years and 10 months she

fell asleep in heightened peace, and in full possession
of her faculties, following to a better life her father Isaac
Herschel, who preceded her after attaining the age of
60 years, 7 months, 17 days and since the 25th March,
1767 lies buried near this spot. *[Der Blick der Verklaerten
war hienieden dem gestirnten Himmel zugewandt; die
eigenen Cometen-Entdeckungen, und die Thielnahme
an den unsterblichen Arbeiten ihres Bruders, Wilhelm
Herschel, zeugen davon bis in die spaeteste Nachwelt.
Die Köningliche Irlaendische Akademie zu Dublin und
die Königliche Astronomische Gesellschaft in London
zaehlten sie su ihren Mitgliedern. In den Alter von 97
Jahren 10 Monaten entschlief sie mit heiterer Ruhe
und bei völliger Geisteskraft, ihrem zu einem besseren
Leben vorangegangen Vater Isaac Herschel folgend der
ein Lebensalter von 60 Jahren, 2 Monathen, 17 Tagen
errichte und seit den 25ten Marz, 1767m hierneben
begraben liegt].*

Caroline began life "with *nothing* but health and abilities
for getting . . . bread." Before she died she had accomplished
a staggering amount of work, including the discovery of eight
comets—five with undisputed priority—as well as several other
celestial objects, as well as the publication of vast volumes
of valuable and vital astronomical information. She had also
received honorary memberships into the most prestigious
scientific societies of her time, and gold medals from royalty
in recognition of her contributions to astronomy.

In 1889, Asteroid 281, which had been discovered the
previous year, was given Caroline's middle name, Lucretia.
In 1935, a lunar crater (in the Sinus Iridium) was named
Caroline Herschel.

In the words of Caroline's fellow astronomer and some-time rival comet-hunter, Mechain, her fame will be held in honor throughout the ages. Though as Caroline herself would want, her name is often mentioned in the same breath as her brother's, like a double star in the night sky, shining most brightly when viewed together.

The William and Caroline statue at the Bath Museum

Timeline

1750 Born on March 16 in Hanover (now Germany).

1754 Recovers from smallpox.

1757 Brother William moves to England; Hanover is occupied by the French, who abandon it the following year.

1761 Contracts typhus.

1767 Father Isaac Herschel dies; eldest brother Jacob Herschel replaces him as the head of the family.

1772 Leaves Hanover for Bath, England; settles down to life in William's home; starts training to be a musician.

1778 Declines a prestigious appointment to sing at the Birmingham Music Festival.

1779 Encouraged by William to start observing stars.

1782 Sings for last time at public concert; William accepts royal appointment as astronomer at Windsor; moves to Datchet with William; starts conducting systematic sweeps of the heavens.

1783 Discovers two nebulae, increasing the number of known nebulae by 2 percent in one night; discovers a total of fourteen new stellar objects by the end of the year; begins recording William's observations of the stars.

1784 Discovers an open cluster in Cygnus NGC 6819.

1786 Discovers comet C/1786 P1 (Herschel) or 1786 II.

1787 Becomes the first professional woman astronomer, when King George III awards her annual lifelong income and the official title of "Assistant Royal Astronomer"; discovers another nebula NGC 7380 in Cephus.

1788 William marries the widow Mary Pitt; discovers comet 35P/Herschel-Rigollet or 1788 II.

1790 Discovers comet C/1790 A1 (Herschel) or 1790 I; discovers comet C/1790 H1 (Herschel) or 1790 III.

1791 Discovers comet C/1791 X1 (Herschel).

1792 Nephew John Herschel is born.

1793 Discovers comet C/1793 S2 (Messier).

1795 Discovers comet 2P/Encke.

1797 Discovers comet C/1797 P1 (Bouvard-Herschel).

1799 The Royal Society publishes her star catalogs under her name (the corrected version of Flamsteed's star Catalog); invited by Maskelyne to the Royal Observatory; begins work on zone catalog.

1804	Completes the first draft of the zone catalog, having recalculated the sweeps and organized 8,760 observations into zones calculated using star coordinates for the year 1800.
1822	William dies; Caroline returns to Hanover to live with brother, Dietrich.
1825	Completes final draft of zone catalog of 2,500 nebulae; organizes data in a format suitable for future observers.
1828	Awarded a Gold Medal by the Royal Astronomical Society.
1835	Elected an honorary member of the Royal Astronomical Society.
1838	Granted an honorary membership of the Royal Irish Academy.
1846	Presented with a gold medal by the king of Prussia.
1848	Dies on January 9 in Hanover.
1889	Asteroid 281, discovered in 1888, is given Caroline's middle name (Lucretia).
1935	A Lunar crater (in the Sinus Iridium) is named Caroline Herschel.

Sources

CHAPTER ONE: The Cinderella of the Family

p. 11, "the street, to make . . ." Michael Hoskin, *Caroline Herschel's Autobiographies* (Cambridge: Science History Publications, 2003), 24.

p. 15, "it was only in the . . . expect to find contentment," Ibid., 43.

p. 17, "more than was necessary for . . ." Mrs. John Herschel, *Memoir and Correspondence of Caroline Herschel*, (London: John Murray, 1879), 20.

p. 17, "the first pair . . . upright finishing the front!" Ibid., 11.

p. 17, "give herself, with all she . . . " Ibid., 136.

p. 19, "threw down his Knife and Fork," Hoskin, *Caroline Herschel's Autobiographies*, 103.

p. 19, "came running and crouched down," Ibid., 26.

p. 21, "only by chance . . . could have been trusted," Ibid., 28.

p. 23, "there was no one who . . ." Ibid., 102.

p. 23, "something like a polished education," Ibid., 37.

p. 23, "against all thoughts of marring . . ." Ibid., 47.

p. 24, "in a Tub of water . . . in the courtyard," Ibid., 36.

p. 24, "doing the drudgery of the scullery," Ibid., 112.

p. 24, "I could not help feeling . . . an Abigail or Housmaid," Ibid., 114.

p. 24–25, "my little Notion of Music . . . " Ibid.

p. 25, "many a wipping," Ibid., 110.

p. 25–26, "Among the group . . . an acquaintance the greater," Ibid., 42.

p. 26, "become a useful singer for . . ." Ibid., 47.

p. 26, "This at first seemed . . . scheme into ridicule," Ibid.

p. 26, "taking in the . . . knew how to sing," Ibid.

p. 26, "knitted as many cotton stockings . . ." Ibid.

p. 28, "had been sacrificed to the . . ." Herschel, *Memoir and Correspondence of Caroline Herschel*, 218.

p. 28, "the Cinderella of the family," Ibid., 299.

CHAPTER TWO: Escape to England

p. 30, "thrown like balls by two sailors," Agnes M. Clerke, *The Herschels and Modern Astronomy* (New York: Macmilliam and Co., 1895), 118.

p. 32, "began imediately giving . . . Gag in my mouth," Hoskin, *Caroline Herschel's Autobiographies*, 119.

p. 33, "we began generally with what . . ." Herschel, *Memoir and Correspondence of Caroline Herschel*, 245-255.

p. 33, "brought home whatever in my . . ." Hoskin, *Caroline Herschel's Autobiographies*, 50.

p. 33, "had the mortification to hear . . ." Ibid., 25.

p. 33, "with so much ill will . . ." Ibid., 120.

p. 33, "Pickpockets and Streetwalkers," Ibid., 121.

p. 33–34, "very seldom I have been . . ." Ibid., 53.

p. 34, "had leisure to try my . . ." Ibid., 50.

p. 34, "began to fear . . . was seldom at home," Ibid.

p. 34–35, "I still was . . . brother," Ibid., 123.

p. 35, "the resolution of never opening . . ." Herschel, *Memoir and Correspondence of Caroline Herschel*, 136.

p. 35, "did me no good, for . . ." Hoskin, *Caroline Herschel's Autobiographies*, 51.

p. 35, "very little better than an idiot," Ibid., 123.

p. 35, "hot-headed old Welsh woman," Ibid., 120.

p. 35, "with a Bason of Milk . . ." Ibid., 122.

p. 35, "an astronomical Lecture of which . . ." Ibid., 51.

p. 36, "was not contented . . ." Ibid.

p. 36, "with making the tube of . . ." Ibid.

p. 39, "much hindered in my practice . . ." Ibid., 122.

p. 39, "to my sorrow I saw . . ." Ibid., 52.

p. 42, "the wicked pilfering wretches by . . ." Ibid., 124.

p. 42, "to drill me for a . . ." Ibid., 55.

CHAPTER THREE: Stars and Song

p. 43, "every leisure moment was eagerly . . ." Hoskin, *Caroline Herschel's Autobiographies*, 54.

p. 45, "was soon brought fainting back . . ." Ibid., 54.

p. 45, "My time was so much taken up . . ." Ibid.

p. XX, "a miscellaneous jumble of elementary . . ." Clerke, *The Herschels and Modern Astronomy*, 121.

p. 47, "I could not help feeling . . ." Ibid., 127.

p. 48, "for speaking my words like . . ." Ibid., 129.

p. 49, "engaged in a long conversation . . .'" Ibid., 131.

p. 50, "never intended to sing any . . ." Ibid., 57.

p. 51, "I have been throughout annoyed . . ." Ibid., 53.

CHAPTER FOUR: Watcher of the Skies

p. 54, "As soon as the public . . ." Hoskin, *Caroline Herschel's Autobiographies*, 62.

p. 54, "Alexander was always . . . have thought a hardship," Ibid., 62.

p. 55, "The mirror was . . . was paid for already," Ibid., 63-64.

p. 55, "Brothers, and the Caster and . . ." Ibid., 64.

p. 57, "opened my mouth for . . ." Ibid., 65.

p. 58, "Never bought Monarch honour so cheap!" Herschel, *Memoir and Correspondence of Caroline Herschel*, 50.

p. 61, "a Telescope adapted for sweeping . . ." Hoskin,
Caroline Herschel's Autobiographies, 71.

p. 62, "sweep for Comets, and by . . ." Ibid.

CHAPTER FIVE: Star Clusters and Mysterious Clouds

p. 72, "My Brother began . . . whole apparatus came
down," Hoskin, *Caroline Herschel's Autobiographies*, 76.

p. 75, "sweeping was interrupted by being . . ." Herschel,
Memoir and Correspondence of Caroline Herschel, 52.

p. 75, "became intirely attached to the . . ." Ibid., 52.

p. 75, "had, however, the comfort to . . ." Hoskin,
Caroline Herschel's Autobiographies, 73.

p. 76, "At each end of . . . my own surgeon," Ibid., 76-77.

p. 76, "till Dr. Lind hearing of . . ." Ibid., 77.

p. 77, "my Brother was no loser . . ." Ibid.

p. 77, "I could give a pretty . . ." Ibid.

p. 77, "personal safety is the last . . ." Ibid.

p. 77, "had a narrow escape of . . ." Herschel, *Memoir
and Correspondence of Caroline Herschel*, 113.

p. 79, "Hier ist wahrhafting ein Loch im Himmel!"
Ibid., 269.

p. 82, "that 2000 would be granted . . ." Ibid., 57.

p. 83, "litigious woman," Ibid., 58.

p. 83, "among all this hurrying business . . ." Ibid., 58.

p. 84, "If it had not . . . of three months" Ibid.

p. 84, "complete workshop for making optical instruments,"
Hoskin, *Caroline Herschel's Autobiographies*, 81

p. 84, "it was a pleasure to go into it," Ibid.

p. 85, "with regrett that . . . being devoured by mice,"
Ibid., 82.

p. 85, "All the Neb are registered . . ." Ibid., 83.

p. 85, "was obliged frequently to sacrifice an hour to her
gossipings," Ibid., 86.

p. 86, "I have calculated 100 nebulae . . . " Ibid., 89.

p. 87, "the object of last night *is a Comet,"* Ibid.

p. 87, "SIR, – In consequence . . ." Herschel, *Memoir and Correspondence of Caroline Herschel*, 65.

p. 87, "excuse the trouble I give . . ." Ibid, 67.

p. 87, "Lastly, I beg of you, sir . . ." Ibid.

CHAPTER SIX: Days Rich in Discovery

p. 89, "I believe the comet has . . ." Herschel, *Memoir and Correspondence of Caroline Herschel*, 69.

p. 89, "give you some account of your comet before I answered it," Ibid., 69.

p. 89, "I am more than pleased . . . it travels very fast," Ibid.

p. 90–91, "200 nebulae of the second thousand," Ibid., 73.

p. 91, "calculated 140 nebulae to-day . . ." Ibid.

p. 92, "before the optical parts were . . ." Ibid., 309.

p. 92, "in a perfect chaos of . . ." Ibid., 73.

p. 92, "very seldom could get a . . ." Ibid., 74.

p. 92–93, "had always some kind of . . ." Ibid., 75.

p. 93, "was one of the nimblest," Ibid., 308.

p. 93, "foremost to get in and . . ." Ibid.

p. 95, "been almost the keeper of . . ." Ibid., 76.

p. 95–96, "dear brother's proposal . . ." Ibid., 178.

p. 96–97, "exactly the sum I saved . . ." Hoskin, *Caroline Herschel's Autobiographies*, 94.

p. 97, "The Cat. of the second . . . 8th of May 1788," Ibid., 95-96.

p. 97–98, "in October I received twelve . . ." Herschel, *Memoir and Correspondence of Caroline Herschel*, 75-76.

CHAPTER SEVEN: Most Admirable Lady Astronomer

p. 100, "his domestic happiness pass into . . ." Herschel,

Memoir and Correspondence of Caroline Herschel, 141.

p. 100, "with saddened heart but unflagging determination,"
Ibid., 141.

p. 100, "I shall be glad to . . ." Constance Lubbock, *The
Herschel Chronicle: The Life-Story of William Herschel
and His Sister Caroline Herschel* (Cambridge:
Cambridge University Press, 1933), 333.

p. 100–101, "Dear Sir – Last night, December 21st . . .
CAROLINA HERSCHEL," Herschel, *Memoir and
Correspondence of Caroline Herschel*, 80.

p. 103, "I am much obliged to . . ." Ibid., 83.

p. 103, "would not affirm that there . . . " Ibid., 82.

p. 105, "did not know what to do," Ibid., 86.

p. 105, "new sweeper," Ibid.

p. 105, "not half finished . . . with my instrument,"
Ibid., 86-87.

p. 105–106, "is a little more . . . not so very rapid," Ibid., 86.

p. 106, "take care to make our . . ." Ibid., 87.

p. 106, "You cannot, my dear Miss . . ." Ibid., 88.

p. 106–107, "second communication, at the same . . ." Ibid., 89.

p. 107, "worthy sister in astronomy," Ibid.

p. 107, "pour . . . illustre que le votre," Ibid., 90.

p. 107, "Savante," Ibid., 89.

p. 107, "Mlle Caroline Herschel, Astronomer Celebre,
Slough," Ibid.

p. 107, "most revered lady," Ibid., 92.

p. 107, "noble and worthy priestess of the new
heavens," Ibid.

p. 107–108, "I still recall the happy hours . . ." Ibid.

p. 108, "in the highest esteem.," Ibid.

p. 111, "could never remember the multiplication . . ." Ibid., 315.

p. 111, "My last paper consisted of . . ." Ibid., 106-107.

CHAPTER EIGHT: Catalog

p. 112, "in petticoats . . . working with little tools," Herschel, *Memoir and Correspondence of Caroline Herschel*, 268.

p. 112, "many a half or whole . . ." Ibid., 259.

p. 114, "happy hours . . . little roof at Slough," Ibid., 332.

p. 114, "Last night, in sweeping over . . ." Ibid., 93-94.

p. 119, "your having thought it worthy . . ." Ibid., 96.

p. 120–121, "Were Flamsteed alive, how . . . are entitled to," Ibid., 103.

p. 121, "little faith in the expedition . . ." Ibid., 94.

p. 121, "came to be detached from the family circle," Ibid., 255.

p. 122, "Uncommonly harassed in consequence of . . ." Ibid., 98-99.

CHAPTER NINE: The Passing of William

p. 123, "I dined at the castle," Herschel, *Memoir and Correspondence of Caroline Herschel*, 113.

p. 123, "The Prince of Orange stepped in . . ." Ibid., 99.

p. 125, "Began to pack up what . . ." Ibid., 105.

p. 125, "ruined in health, spirit, and fortune," Ibid., 116.

p. 126, "according the old Hanoverian custom . . ." Ibid.

p. 126, "the time I bestowed on . . ." Ibid.

p. 128, "was wanted to assist my . . ." Ibid., 120.

p. 128, "every shelf and drawer," Ibid., 130.

p. 128–129, "found on looking . . . of the night," Ibid., 129-130.

p. 129, "LINA, - There is a great comet . . ." Ibid., 131.

p. 129, "I keep this as a relic . . ." Ibid.

p. 129, "Here closed my Day-book . . ." Ibid., 132-133.

p. 129, "seized by a bilious fever," Ibid., 135.

p. 129, "rise for a few hours . . ." Ibid.

p. 129–130, "I was entirely confined . . . happen to me,"
Ibid., 135-136.

p. 130, "not one comfort was left . . ." Ibid., 138.

p. 130, "acquire fortitude enough . . . this world," Ibid.,
133-134.

p. 130, "at that time I should . . ." Ibid., 200.

p. 131, "from all these sorrowing friends . . ." Ibid., 139.

CHAPTER TEN: Why Did I Leave Happy England?

p. 132, "I saw no one I . . ." Herschel, *Memoir and
Correspondence of Caroline Herschel*, 139.

p. 132, "the last hope of finding . . ." Ibid., 219.

p. 132, "whole life almost has passed . . ." Ibid., 136.

p. 133, "abominable city," Ibid., 247.

p. 133–134, "What little I have seen . . ." Ibid., 156.

p. 134, "now quite a new world . . ." Ibid., 215.

p. 134, "From the first moment I . . ." Ibid., 200.

p. 134, "I found I was alone," Ibid.

p. 135, "the handsomest rooms, three or . . ." Ibid., 155.

p. 135, "hard to live for months . . ." Ibid., 156.

p. 135 136, "comparing the country in which . . ." Ibid., 238.

p. 136, "Believe me, I would not . . ." Ibid., 163.

p. 137, "learned society, or blue-stocking club," Ibid., 192.

p. 137, "honoured with a wie gehts? by His Majesty,"
Ibid., 303.

p. 137, "always sure to be noticed . . ." Ibid., 276.

p. 139, "amusing herself with having the . . ." Ibid., 171.

CHAPTER ELEVEN: Double Stars

p. 140, "I wish to live a little longer," Herschel, *Memoir and
Correspondence of Caroline Herschel*, 171.

p. 140–141, "that I might make you . . ." Ibid.

p. 142, "prize, and more than prize . . ." Ibid., 188.

p. 142, "learned fully to appreciate the . . ." Clerke, *The Herschels and Modern Astronomy,* 132.

p. 142–143, "recall the health, eyesight, and . . ." Herschel, *Memoir and Correspondence of Caroline Herschel*, 196.

p. 143, "find undoubtedly many . . . pretty numerous catalog," Ibid., 197.

p. 144, "was none of my doings," Ibid., 227.

p. 144, "the completion of a series . . ." Ibid., 223-224.

p. 144, "Miss Herschel it . . . industry of the sister," Ibid., 224.

p. 144, "clumsy," Ibid., 232.

p. 145, "more shocked than gratified by . . ." Ibid., 231.

p. 145, "Whoever says *too much of* . . ." Ibid., 142.

p. 145, "he did all," Ibid., 206.

p. 145, "I did nothing for my . . ." Ibid., 142.

p. 145, "I am nothing, I have . . ." Ibid., ix.

p. 145, "meet with a good- natured . . ." Ibid., 199.

p. 145–146, "wonderfully well and very nicely . . ." Ibid., 254-255.

p. 146, "Ja! If I was thirty . . ." Ibid., 256.

p. 146, "jingled glasses with Betty," Ibid., 257.

p. 146, "entire confidence and approbation," Ibid., 340.

p. 146, "Es lebe Sir John! Hoch! Hurrah!" Ibid., 257.

p. 146, "in idea every inch," Ibid., 177.

p. 146, "As soon as your instrument . . ." Ibid., 258.

p. 146, "It is not *clusters of* . . ." Ibid., 269.

p. 146, "for that does not answer . . ." Ibid.

p. 147, "something more than a total . . ." Ibid., 258.

p. 147, "I have swept well over Scorpio," Ibid., 270.

p. 147–148, "many entries in my sweeping . . ." Ibid.

p. 149, "could not help lamenting," Ibid., 247.

p. 149, "could not take to his . . ." Ibid.

p. 149, "completely happy," Ibid.

p. 149, "the tests of astronomical merit . . ." Ibid., 227.
p. 149, "I cannot help crying aloud . . ." Ibid., 301.
p. 149, "I think it is almost . . ." Ibid., 313.
p. 150, "I am the only one . . ." Ibid., 168.
p. 150, "Coming to *me* with such . . ." Ibid., 285.
p. 151, "Ach mein Gott! das ist . . ." Ibid., 295.
p. 151, "as if we had parted . . ." Ibid.
p. 151, "Next to listening to the . . ." Ibid., 252.
p. 151, "nimbleness in walking," Ibid., 237.
p. 151, "the good folks are not . . ." Ibid.
p. 151, "the few, few stars I . . ." Ibid., 216.
p. 151–152, "melancholy lethargy," Ibid.,264.
p. 152, "store of astronomical and other . . ." Ibid., 264.
p. 152, "all that might produce nonsense . . ." Ibid.
p. 152, "I would not wish on . . ." Ibid., 323-324.
p. 152, "to be remembered . . . for yet awhile," Ibid., 301.
p. 152, "little gratification," Ibid., 337.
p. 152, "immortal brother, Sir William Herschel," Ibid.
p. 153, "breathed her last at eleven . . ." Ibid., 344.
p. 153–154, *Der Blick der Verklaerten war. . .*" Ibid., 351.
p. 154, "with *nothing* but health and . . ." Ibid., 170.

Bibliography

Apfel, N. H. *Nebulae: The Birth and Death of Stars*. New York: Lothrop, Lee, and Shepard Brooks, 1988.

Ashton, H. and K. Davies. *I Had a Sister: A Study of Mary Lamb, Dorothy Wordsworth, Caroline Herschel, Cassandra Austen*. London: Lovat Dickson Limited, 1937.

Betham-Edwards, M. *Caroline Herschel: Astronomer, Mathematician*. London: Griffith and Farran, 1884.

Branley, F. M. *Comets*. New York: Thomas Y. Crowell, 1984.

Brockhaus Enzyklopädie. *Caroline Herschel*. Mannheim: Brockhaus, 1987.

Buttmann, G. *Wilhelm Herschel, Leben und Werk. [Große Naturforscher, Band 24]*. Stuttgart: Wissenschaftliche Verlagsgesellschaft m.b.H., 1961.

Clerke, Agnes M. *The Herschels and Modern Astronomy*. New York: Macmilliam and Co., 1895.

Cooney, M.P., ed. *Celebrating Women in Mathematics and Science*. Reston: National Council of Teachers of Mathematics, 1996.

Corkran, A. *The Romance of Woman's Influence: St. Monica, Vittoria Colonna, Madame Guyon, Caroline Herschel, Mary Unwin, Dorothy Wordsworth and other mothers, wives, sisters, and friends who have helped great men*. Glasgow: Blackie and Son, 1912.

Feyl, R. *Caroline Herschel (1750-1848): Aufbruch in die nicht gewollte Selbständigkei; Bedeutende Frauen Hannovers.* Hannover: Sophie& Co., 1991.

Gallant, R. A. *Comets, Asteroids, and Meteors.* New York: Marshall Cavendish, 2001.

Gingerich, O. "Through Rugged Ways to the Galaxies." *Journal for the History of Astronomy* 21, (1990): 77-88.

Harris, N. *Space.* San Diego: Blackbirch, 2002.

Herschel, C. "Epitaph." *Scripta Mathematica* 21, (1955): 251.

Herschel, Mrs. J., ed. *Memoirs and Correspondence of Caroline Herschel.* London: John Murray, 1876.

Herschel, J. F. W. "General Catalog of Nebulae and Clusters of Stars." *Philosophical Transactions of the Royal Society* 154, (1864): 1-137.

Herschel, J. F. W. *A Preliminary Discourse on the Study of Natural Philosophy [Worlds of Desire: The Chicago Series on Sexuality, Gender, & Culture]* Chicago: University of Chicago Press, 1987.

Higgins, F. L. *Sweeper of the Skies: A Story of theLlife of Caroline Herschel, Astronomer.* Chicago: Follett Publishing Company, 1967.

Hoskin, M. *Caroline Herschel's Autobiographies,* Cambridge: Science History Publications, 2003.

———. "Caroline Herschel: Assistant Astronomer or Astronomical Assistant?" History of Science 40, no. 4 (2002): 425-444.

———. *The Herschel Partnership: As Viewed by Caroline.* Cambridge: Science History Publications Ltd., 2003.

———. "The Real Caroline Herschel." *Bulletin of the American Astronomical Society* 35, (2003).

Hughes, D. W., "Caroline Lucretia Herschel: Comet Huntress." *Journal of the British Astronomical Association* 109, (1999), p. 78.

Hunter, L., and S. Hutton, eds. *Women, Science and Medicine 1500-1700*. Gloucestershire: Sutton Publishing, 1997.

Jones, J. H. "The Caroline Herschel Objects." *Sky and Telescope* 104, no. 5 (2002), 107-111.

Kempis, M.T.A. "Caroline Herschel." *Scripta Mathematica* 21, no. 4 (1955): 237-251.

Lada, C. J., "Vacancies in the Heavens: Caroline Herschel and the discovery of dark nebulae." *Bulletin of the American Astronomical Society* 36, (2004): 1413.

Lubbock, C. A., ed. *The Herschel Chronicle: The Life-Story of William Herschel and His Sister Caroline Herschel.* Cambridge: Cambridge University Press, 1933.

Mädler, J. H. *Geschichte der Himmelskunde*. Braunschweig: Westermann, 1873.

Moore, Patrick. *Caroline Herschel: Reflected Glory*. Bath: William Herschel Society, 1988.

O'Dell, C. R. *The Orion Nebula: where stars are born.* Cambridge: Harvard University Press, 2003.

Ogilvie, M B. "Caroline Herschel's contributions to astronomy." *Annals of Science* 32 (1975): 149-161.

Osen, L. *Women in Mathematics*. Cambridge: MIT Press, 1974.

Prinja, R. *Comets, asteroids, and meteors*. Chicago: Heinemann, 2002.

Robinson, E. M.W. *Stars in her heart*. New York: TEACH Services Inc., 2005.

Ruskin, S. *John Herschel's Cape Voyage: Private Science, Public Imaginations, and Ambitions of Empire [Science, Technology, and Culture, 1700-1945].* Burlington: Ashgate Publishing, 2004.

Sadie, J. A. and S. Stanley. *Calling on the Composer: A Guide to European Composer Houses and Museums.* Yale: Yale University Press, 2005.

Schad, M. *Frauen die die Welt bewegten*. Munich: Pattloch Verlag gmbH und Co. KG, 2000.

Seymour, S. *Comets, Meteors, and Asteroids*. New York: William Morrow and Co., Inc., 1994.

Soter, S. and N. Tyson, eds. *Cosmic Horizons: Astronomy at the Cutting Edge*. New York: New Press, 2001.

South, J. An address delivered at the annual general meeting of the Astronomical Society of London on February 8, 1829, on presenting the honorary medal to Miss Caroline Herschel. *Memoirs of the Royal Astronomical Society* 3 (1829): 409-412.

Stott, C. *Caroline Herschel. Into the Unknown [Women History Makers]*. London: Macdonald, 1988.

Strohmeier, R. *Lexikon der Naturwissenschaftlerinnen und naturkundigen Frauen Europas: Von der Antike bis zum 20. Jahrhundert*. Frankfurt am Main: Verlag Harri Deutsch, 1998.

Sylvester, C. *Feminist International Relations: An Unfinished Journey*. Cambridge: Cambridge University Press, 2002.

Tabor, M. E. *Pioneer Women*. London: The Sheldon Press, 1921.

Teresi, D. *Lost Discoveries*. New York: Simon and Schuster, 2002.

Vogt, G. L. *Asteroids, Comets and Meteors*. Brookfield: The Millbrook Press, 1996.

Vordermaan, C. *How Maths Works*. London: Dorling Kindersley Limited, 1996.

Yeomans, Donald K. *Comets: A Chronological History of Observations, Science, Myth, and Folklore*. New York: John Wiley and Son, 1991.

Web sites

MacTutor History of Mathematics Archive of the School of Mathematics and Statistics, University of St. Andrews, Scotland

http://www-history.mcs.st-andrews.ac.uk/Mathematicians/Herschel_Caroline.html

Bath Preservation Trust/William Herschel Museum

http://www.bath-preservation-trust.org.uk/museums/herschel/

Agnes Scott College

http://www.agnesscott.edu/lriddle/women/herschel.htm

StarChild. A site maintained by the High Energy Astrophysics Science Archive Research Center and Dr. Nicholas E. White (Director), within the Astrophysics Science Division at NASA.

http://starchild.gsfc.nasa.gov/docs/StarChild/whos_who_level2/herschel.html

Astronomical Society of the Pacific

http://www.astrosociety.org/education/resources/womenast_bibprint.html#herschel

Index